STEAM start

KINDERGARTEN

Adventures in Shapes and Space

Units 1-5
Science, Technology, Engineering Design, Art, Math

STEAMStart
By Jeannie Ruiz

Printing FutureS

STEAMStart: A Year of Hands-On Discovery

Kindergarten Adventures in Shapes & Space Units 1-5

Published by Printing Futures Vancouver, WA

Author: Jeannie Ruiz Editor: Wayne Stripling, Jr.

Edition: 2025

ISBN 978-1-942357-23-0 Hardcover

Photo permissions and authors, illustrators, designers, photographers available for download at the URL.

http://www.insl.org/STEAMSTART

STEAMSTART DEVELOPMENT TEAM
STEMWORKS EXEMPLARY PROGRAMS DEVELOPED BY Art of STEM and The International STEM League, Inc.

(Author, Jeannie Ruiz)

Art of STEM developed STEAMStart as well as upper grade level programs that are included in the STEMWorks Database by Change the Equation, a non-partisan, CEO-led initiative to connect and align efforts to improve STEM learning.

STEMWORKS
CERTIFIED
READY-TO-SCALE

RECOMMENDED BY REAL TEACHERS

We are thoroughly enjoying our STEAM lessons. The program is excellent! My students have loved making the Rhombis and getting to build with shapes and tape. It has been a big hit! Thanks for everything!
Alicia Stenard
Mater Christi in Albany School District

"We need more projects like this that carry on for a longer number of weeks. My teachers used scope and sequence suggestions, and the program ran after school for a semester. We could have worked on this a lot longer."
Karen James
Druid Hills Elementary in Charlotte, NC

"I'm excited for our students to get their hands on real STEAM projects. There isn't much out there for Kindergarten and First Grade."
Sandy Mettler
Fort Worth Schools in Dallas, TX

KINDERGARTEN ADVENTURES IN SPACE AND SHAPE
AUTHOR'S COMMENTS

Designers collect, analyze, and synthesize data guided by specifications. Designers make clear and concise recommendations using drawings,verbal descriptions, and math models.

The Industrial Designers Society of America's definition of industrial design describes design as "the professional service of creating and developing concepts and specifications that optimize the function, value and appearance of products and systems for the mutual benefit of both user and manufacturer."

Rhombi asks her friends to offer assistance in design. Students collect, analyze and make data-based decisions. They must share their concepts verbally, using drawings, and through models (scale and math).

First of all, thank you for taking the leap. Project-based learning requires a commitment on the part of the teacher. It can be messy. It can mean more planning and more time spent gathering odds and ends that make a project work. The time and energy you spend developing imaginative responses and ability to problem solve will pay off tenfold.

There is significant evidence in our own research that indicates a need for memorization and rote learning IN ADDITION TO LONG-RANGE PROJECTS and innovative problem-solving. A decent project uses the challenge as a springboard for planning skills-based lessons. A great project urges kids to seek skills and knowledge that you can offer because they want to succeed in solving the problem you've posed. Quick recall of facts is necessary for innovative thinking. Kids don't need to stop and look up simple facts or count on teddy bears when we want them to focus on collecting data and making reasonable decisions.

Projects need outcomes. They need number-driven goals that qualify success in the challenges. Let kids get creative and try things that will fail. Did I say to let them fail? Let them fail. Innovators fail all the time. Some of our greatest inventions are the result of absolute and profound failures. Learning to use the failure as a tool rather than an emotional trial is where students are headed. It takes a few falls before learning to ride a bike. They will make a few bouncy balls on the road to a great slimy wall-walker.

The question arises as to why we chose to call students "designers" rather than "engineers" or "scientists." Designers incorporate a wide variety of communication skills as well as industry expertise. Because communication of ideas among students is integral to the success of any project-based curriculum, the students become Rhombi's designers.

The Art of STEM Team celebrates the amazing lessons we learn from you. When you develop a new technique or unique approach to the material, let us hear from you. The resource site grows stronger as teachers share their ideas with each other.

Sincerely

,

Jeannie Ruiz
President of Ten80
Elementary

ADVENTURES IN SPACE AND SHAPE
CONTENTS

ADVENTURES IN SPACE AND SHAPE
CONTENTS

RHOMBI'S ADVENTURES IN SPACE AND SHAPE

Learning shapes is great. Doing short-term interactive projects that involve shapes is even better. Turning those projects into inquiry-based challenges takes things a step further. Involving students in long-range projects that require multiple steps, depth of knowledge, and true integration of science and math is the goal. The National Science Foundation funded Innovative Technology Experiences for Teachers (ITEST) research that developed a list of criteria for excellence in project based learning.

CRITERIA FOR EXCELLENCE IN PROJECT BASED LEARNING

- Focus on activities that motivate students to develop interest and skills in STEM careers.
- Provide intensive engineering experience of a long duration (academic year and/or summer proportional to the nature of the program).
- Use technology tools and hands-on learning.
- Involve industry mentors, scientists, and engineers.
- Incorporate students' self-regulated learning within specific engineering design experiences.
- Offer innovative, high interest STEM activities.
- Involve participants in the designing and building of artifacts.
- Use scientific integration and/or the engineering design process.
- Engage students in project-based and/ or problem-based learning.
- Develop participants' technical skills (as appropriate to the project).
- Make explicit connections between (STEM) academics and project.
- Build in success experiences and opportunities to learn from failure.
- Explore technical careers from technicians to engineers and engineering management.

RHOMBI'S ADVENTURES IN SPACE AND SHAPE
GEOMETRY IS THE THREAD

The word "geometry" comes from the Greek words for "Earth" and "measure." Geometry was first used to measure and chart the length, area and shape of land surfaces.

GEOMETRY is the single math thread that runs through Common Core from Kindergarten to 12th grade. STEAMStart promotes STEAM and the Art of STEM through an authentic experience in engineering problem solving built on a platform of geometric constructions, analytical geometry, and engineering design. Math is a gateway to success in professional fields: speaking the language of mathematics fluently means that students may define their career paths by choice, not resignation.

~ Jeannie Ruiz

STEAMStart Developer

THE MATH GENE
Contrary to popular belief, no one is born with a genetic predisposition toward math. There is no "math gene" in humans (Keith Devlin, The Math Gene); mathematics is a learned skill for everyone just like basketball, playing the French Horn, or ballet. Humans have brains which are already wired to enable math acquisition (Stanislas Dehaene, Number Sense 1997). Infants, chimps, dogs, and birds can recognize groups of objects up to three and decide that an object has been taken away or added. That is not "doing math," but it is a qualitative foundation on which to build understanding. Math instruction in school develops the initial qualitative sense of space, time, and quantity into the descriptive -and highly useful-language of mathematics. Children should learn all things in context.

MULTI YEAR FLUIDITY
Geometry as a foundation allows the educator to implement multi-year projects that students enhance year after year. The essence of the project is still fluid, while monthly, weekly, and daily objectives are met for the grade level.

NEXTGEN SCIENCE STANDARDS - STRUCTURE AND FUNCTION
Engineering and Design: Students should understand that "the shape and stability of structures of natural and designed objects are related to their function(s). (K-2-ETS1-2)."

GREAT S.T.E.A.M. TEACHER CHECKLIST

1. DO MORE THAN IMPART KNOWLEDGE Children will teach themselves if offered the opportunities to explore and problem solve. Teach by posing questions. Teach by developing curiosity. Teach by guiding learnerswithinteractiveexperiences.Teach to learn. Learn as much as you teach. Adults often take so much control of the teaching process that children resort to mimicry as a way of pleasing their mentors. Every child learns at a unique pace and with a unique set of circumstances; therefore each child requires a unique path. If educators view learning as a problem-solving process, learning becomes an individual journey.

2. DISLODGE THE MISCONCEPTIONS Every child enters school with a unique set of skills and preconceptions about learning. Every child has already formed opinions based on observation, mimicry, and the outcome of challenges (how were successes and failures handled by adults). Some facts that lodge in a child's mind are erroneous. Other ideas are keen observations of the natural and technological world. Facilitate learning. Increase knowledge. Guide inquiry. Open minds to the possibilities. Offer multiple opportunities to learn and approaches to problem solving as a way of addressing misconceptions.

3. USE WHOLE GROUP SPARINGLY There are rare teachers with the ability to engage learners as a group. This interaction has a fast-paced Q&A quality and -generally- involves a points system or prizes. The up-side is that learners are involved as a group and learn courteous behavior. The down-side is that educators are rarely able to get individual feedback or offer reinforcement / stimulation. Even technologically based feedback systems are an impersonal way to discover a student's strong and weak areas.

4. LEARN MORE WITH SMALL GROUPS Teachers that interact in small groups with students on a regular basis know far in advance of testing exactly what the outcomes will be. Anxiety lessens for teachers and learners. Differentiation becomes possible, and students no longer "tune out" the instructor when the instructor becomes a facilitator. It's hard to ignore the teacher when she's sitting at your table asking questions and challenging assumptions. It's difficult to avoid detection if the investigation requires active participation. For the learner, there is no more "tuning out" or sleep-walking through the day.

5. ASSESS STUDENTS REGULARLY There are no surprises when learners take part in authentic and active assessments. Great teachers know well ahead of time how students will perform on standardized tests.

CONFIDENT S.T.E.A.M. STUDENTS

1. They look forward to doing science.
2. They demonstrate a desire to learn more.
3. They seek to collaborate and work cooperatively with their peers.
4. They are confident in doing science;
   they demonstrate a willingness to modify ideas, take risks, and display healthy skepticism.

1. Children exhibit curiosity and ponder observations.
2. They move around selecting and using the materials they need.
3. They take the opportunity and the time to "try out" their own ideas.

1. Children design a way to try out their ideas, not expecting to be told what to do.
2. They plan ways to verify, extend or discard ideas.
3. They carry out investigations by: handling materials, observing, measuring, and recording data.

1. Children express ideas in a variety of ways: journals, reporting out, drawing, graphing, charting, etc.
2. They listen, speak and write about science with parents, teachers and peers.
3. They use the language of the processes of science.
4. They communicate their level of understanding of concepts that they have developed to date. Children Propose Explanations and Solutions and Build a Store of Concepts.

1. They use investigations to satisfy their own questions.
2. They sort out information and decide what is important.
3. They are willing to revise explanations as they gain new knowledge.

1. Children ask questions (verbally or through actions).
2. They use questions to lead them to investigations that generate further questions or ideas.
3. Children value and enjoy asking questions as an important part of science.

1. Children observe, as opposed to just looking.
2. They see details, they detect sequences and events; they notice change, similarities and differences, etc.
3. They make connections to previously held ideas.

1. They use indicators to assess their own work
2. They report their strengths and weaknesses.
3. They reflect with their peers.

"Inquiry Based Science: What Does It Look Like?" Connect Magazine (published by Synergy Learning), March-April 1995, p. 13.

ADULT MISCONCEPTIONS

These misconceptions were identified based on in-depth interviews with early childhood teachers about the key issues in early mathematics education as well as researchers' experiences in teaching early childhood students, conducting workshops with early childhood teachers (Ginsburg, Jang, Preston, Van Esselstyn & Appel, 2004; Ginsburg et al., 2006), working with them in early childhood classrooms, and engaging in informal conversations with them. The study also based descriptions of the myths on available research literature (Ginsburg, Lee & Boyd, 2008). The nine misconceptions defined by Joon Sun Lee of Hunter College, The City University of New York and Herbert P. Ginsburg of Teachers College, Columbia University are:

1. Young children are not ready for mathematics education.
2. Mathematics is for some bright kids with mathematics genes.
3. Simple numbers and shapes are enough.
4. Language and literacy are more important than mathematics.
5. Teachers should provide an enriched physical environment, step back, and let the children play.
6. Mathematics should not be taught as stand-alone subject matter.
7. Assessment in mathematics is irrelevant when it comes to young children.
8. Children learn mathematics only by interacting with concrete objects.
9. Computers are inappropriate for the teaching and learning of mathematics.

These misconceptions often interfere with understanding and interpreting the new recommendations on sound early childhood mathematics education, and become subtle (and sometimes overt) obstacles to implementing the new practices in the classrooms (Richardson, 1996).

DEVELOPING A S.T.E.A.M. MINDSET

While STEAM (Science, Technology, Engineering, Arts, Mathematics combined) is a new buzz word in the field of Education, it was already a well established practice before Mozart gave his first concert as a child prodigy (Salzburg, Austria, 1761).

Learning to incorporate math modeling, grounding the symbolic nature of numbers in reality, and turning the language of mathematics into an integral part of ALL subjects are the true keys to successful STEAM classrooms. Developing confidence in the understanding of size, scale, number, place value, data collection, data analysis and real problem-solving reasonableness is necessary for the creation of a true STEAM culture.

Start each day with a measurement and estimation challenge. As students interact with blocks, toys, manipulatives and other classroom objects, connections should be drawn to their counterparts in the real world.

ex: Oh, Thomas, that is a 3ft tall tower you've built. Did you know that the Tower of London is 89ft tall? You could tell us a really great story about that tower. Renoir painted towers. Let's look at some of his online. What shapes do you see in the images? Maybe you could paint a Renoir style tower to illustrate your story about the scale model 3ft tower you've designed. If we had more blocks, what else would you add to the tower? A drawbridge? Wow, let's look at the design of a drawbridge's pulley system.

Expecting that level of interaction is realistic when learning centers are guided by the teacher. If all students will be using blocks, you can bet they'll build towers. Anticipating student interests allows the teacher to prepare for verbal interactions that lead to STEAM experiences.

Project-based learning dictates that teachers take the long view in planning. Begin with what you need to accomplish in order to meet some challenge (like addressing Common Core and Next Generation standards). What project will allow natural extensions to your subject matter but also draw kids in as far as interest level? Making the project tie in to real life is also advisable. Do you like to quilt? Incorporate that into your classroom as a tie to measurement, geometry, patterns, history, writing, etc. The project might be designing and making quilts for local shelters. Get creative. In the case of STEAMStart, we've given you the jumpstart. Now, you get to extend the project to other aspects of learning in your curriculum.

DEFINING S.T.E.A.M.

STEAM (science, technology, engineering, arts, mathematics) is just now becoming an accepted acronym among teachers and is seen as a somewhat mysterious -almost magical- fix for all the problems American schools and businesses are facing. Perhaps we should start with what STEAM is not:

- STEAM is not a conglomeration of lessons culled from five subject areas.
- STEAM is not something you can cover in a unit or two every year.
- STEAM is not a 5-day or 10-day curriculum piece.
- STEAM cannot be delivered in the traditional pencil-and-paper method by a sage on the stage standing behind the classroom lectern or seated behind a grand old wooden desk. There are no "Buehler..?" moments in a STEAM environment. *(reference the movie Ferris Buehler's Day Out)*

STEAM is a philosophy of teaching, learning and working. STEAM is more than just the sum of the words in its acronym.

In other words, do real stuff to find real answers to real problems.

What has become a raging debate among educators seems pretty simple to the professionals that actually work in STEAM fields. Academicians will spout research from their PhD thesis and published articles in journals, etc. Ask a scientist, a mathematician, an engineer or a programmer how to prepare kids for a STEAM workplace and they'll tell you to make kids DO something.

While activity centers are typical in kindergarten settings, they have all but disappeared by second grade. Bring them back. Small group interaction is optimal for STEAMStart as it is for any STEAM program. Why do you want to use stations or centers in the classroom?

- Simplifies planning - plan over a week rather than day by day...
- Offers opportunities for more cohesive lessons...
- Allows teachers to hold small group sessions daily...
- Encourages teamwork...
- Simplifies classroom management... *
- STEAM increases the ability to draw connections among various concepts.

* reference Harry Wong and https://www.effectiveteaching.com

BUILDING S.T.E.A.M. CLASSROOMS

According to corestandards.org, "The Common Core State Standards provide a consistent, clear understanding of what students are expected to learn, so teachers and parents know what they need to do to help them. The standards are designed to be robust and relevant to the real world, reflecting the knowledge and skills that our young people need for success in college and careers. With American students fully prepared for the future, our communities will be best positioned to compete successfully in the global economy."

Support Common Core Standards and Learning Goals

Support the ongoing STEAM project with a permanent station and construction materials. Recycled and up-cycled materials should be available for use along with tools, paper, pencil and data collection opportunities.

Encourage discussion of STEAM subjects through the arts as well as investigations involving art inventions, art history, genres, musical instruments and sound, etc. Create a "stage" where students give skits, give presentations, teach lessons and make up songs about classroom learning.

Keep blocks available all the time along with small scale toys at different scales. Add books about world architecture, habitats, and building projects.

Facilitate the transition from fiction and storybooks to gleaning facts from informational text. Fill your library with fiction AND non-fiction. Encourage students to write informational text and to keep journals for science and math.

Most kindergarten classrooms include an area for dress-up and pretend. How many first and second graders would still benefit from this type of role-play? Shoot for higher level characterization with lab coats, construction hats, medical scrubs, artist's smocks, etc.

LONG RANGE PROJECT

If we had a floating house (school, city, state) that could go anywhere on the ocean, where would we want to be?

Instead of being planted in one place, what if we could take our house (school, city, state) with us? Turtles do. No matter where the turtle goes, it's always home.

Will we stay in the same place all the time, or move with currents, weather patterns and seasons?

Build a room, house, school, city, state, or country as a floating island.
Basic elements are squares and triangles which combine (add) to form a solid. Students begin to seek skill in STEAM subjects as a way to solve problems and overcome challenges... as long as the project is interesting enough to keep their attention.

Begin investigating:

- How much milk is delivered to your school everyday?
- Where does it come from?
- What is the source of milk on the floating city-state island?
- How many people can live in a defined habitat? What do they require for survival and comfort?
- How will you move from island to land?
- What happens to structures in various kinds of weather?
- Can we stay out of the way of huge storms over the ocean?
- Do we want to move with the seasons?
- Will it move? How fast do things move?
- If we move, what will fuel the island?
- How much will all that cost, by the way?
- Where is the funding source?

and on... and on... and on...

Since we can't build full-size during the planning phase, we'd better start with scale models. The models will save time, energy, materials, cost and headache. Let's start with the basic information on geometry, polygons and structures.

Introducing STEAMStart.
"Once upon a time, there was Rhombi. She loved shapes..."

USING THE MATERIALS

ENGAGE Each activity uses a read-aloud book to jumpstart the investigation and engage students.

DISCUSS Establish mindbugs with an introductory discussion for each investigation. The mindBug is a misconception in student thinking; procedural understanding that worked in one situation but is incorrectly applied to other skills. Like a computer virus, mindbugs spread through a student's system.

Children often think that air weighs nothing because they see scales measuring zero from an early age.

Example in Math:
Students line the numbers up on the right to add. This works until 5th grade when they attempt to add decimals.

EXPLAIN Each activity includes a brief background for the purpose of introducing the subject matter.

MOVE Get students up and moving to activate that extremely useful muscle memory.

INVESTIGATE

Although each EXPLORE activity and investigation stands alone, together they also build knowledge required to meet the long-range project. In Module 1, students will -ultimately- be challenged to help Rhombi prepare for a trip.

ELA Build skills in communication, writing and speaking.

UNIT 1
CIRCLE TIME

CIRCLE TIME

Essential Questions

1. What is a way to prove that a shape is a circle and not just an oval?
2. Can you talk about shapes using attributes?
3. How would you decide which wheel to use on a bike, cart or skate?
4. What are you trying to accomplish in Rhombi's Challenge?

ENDURING UNDERSTANDING

"Halves" means two equal parts." Circles and rectangles break into equal parts.

Breaking circles or rectangles into more equal parts means that the parts will be smaller.

Three objects may be put in order from longest to shortest by comparing their lengths.

The length of an object may be discussed using whole numbers.

A smaller object may be used as a measurement tool when measuring longer lengths.

Organizing information helps in discussion.

Focus on circles and things that move in circles. Look for circles, and begin a long-range weather study. Create sturdy pinwheels and meaningful circle trees. Explore radius and diameter via wingspan.

Students will apply basic measurement, understanding of number, place value, data collection, and data application in problem solving.

Rhombi wants to give her uncle a gift that rotates or revolves. Students will design and create a spinning or looping piece of art, toy, or structure. Examples may include a pinwheel, anything that hangs from a single ribbon, pine cone bird feeders, or ornaments.

designer

one who plans (usually in graphic form) and presents plans to execute new ideas

rotate

spin around a central axis

CIRCLE TIME
CURRICULUM CONNECTIONS

Rhombi wants to give her uncle a spinning (rotating) toy.

MATH BACKGROUND - Circles

A circle is the set of points in a plane that are equidistant from a given point.

Radius is the distance from center. Diameter is twice the radius. From its center, the circle's angle extends to a full 360 degrees.

TECHNOLOGY CONNECTIONS

Visit the blog for links to excellent S.T.E.A.M. geometry activities and investigations.

- Geometry Games for Kids
- Book Lists
- Links to Apps

HISTORICAL CONNECTIONS

Natural circles like the Sun and Moon exist.

Evidence of wheeled vehicles appears from the mid-4th millennium BC in the Neolithic age.

The first recorded algorithm for Pi was developed in 250 BC by the Greek mathematician Archimedes**.

* http://en.wikipedia.org/wiki/Circle#History
** http://en.wikipedia.org/wiki/Pi#History

ART CONNECTIONS

Artists have used circles since earliest recorded history. From mandalas to crop circles, circles appear in all cultural art. Modern artists present the greatest opportunity for class experiences.

Joan Miro was a Spanish painter, sculptor, and ceramicist (1893-1983). GOOGLE: Joan Miro Art

Circles in a Circle from 1923 by Vasily Kandinsky, (1866 - 1944) is an oil on canvas. GOOGLE: Vassily Kandinsky circles in a circle

Download free mandala patterns. GOOGLE: free mandala patterns

GREAT READING CONNECTIONS

Circle Unbroken
By Margot Theis Raven

So Many Circles, So Many Squares By Tana Hoban

Pond Circle
By Betsy Franco

Pumpkin Circle: The Story of a Garden George Levenson

CIRCLE TIME
CURRICULUM CONNECTIONS

COMMON CORE CONNECTIONS

ELA/Literacy

W.K.7 Participate in shared research and writing projects (e.g., explore books by a favorite author and express opinions). (K-ESS2-1)

Mathematics

MP.2 Reason abstractly and quantitatively. (K-ESS2-1)
MP.4 Model with mathematics. (K-ESS2-1)
K.CC.A Know number names and the count sequence. (K-ESS2-1)
K.MD.A.1 Describe measurable attributes of objects, such as length or weight. Describe several measurable attributes of a single object. (K-ESS2-1)
K.MD.B.3 Classify objects into given categories; count the number of objects in each category and sort the categories by count. (K-ESS2-1)

NEXT GEN SCI CONNECTIONS

K-2-ETS1-1. Ask questions, make observations, and gather information about a situation people want to change to define a simple problem that can be solved through the development of a new or improved object or tool.
K-2-ETS1-2. Develop a simple sketch, drawing, or physical model to illustrate how the shape of an object helps it function as needed to solve a given problem.
K-2-ETS1-3. Analyze data from tests of two objects designed to solve the same problem to compare the strengths and weaknesses of how each performs.
K-PS2-1. Plan and conduct an investigation to compare the effects of different strengths or different directions of pushes and pulls on the motion of an object.
K-PS2-2. Analyze data to determine if a design solution works as intended to change the speed or direction of an object with a push or a pull.*
1-ESS1-1. Use observations of the sun, moon, and stars to describe patterns that can be predicted.

UNIT 1 S.T.E.A.M. ACTIVITIES FOR “CIRCLE TIME”

UNIT 1: Circle Time: Science Activity: Begin documenting wind and
UNIT 1: Circle Time: weather. Technology Activity: Create a counting tool.
UNIT 1: Circle Time: Engineering Activity: Decide what material works best for a
UNIT 1: Circle Time: pinwheel. Art & Music Activity: Create a classroom “Circle Tree.”

UNIT 1: Circle Time: ELA Activity: Read “Circle Time” aloud. Students illustrate the story.

CIRCLE TIME
BUILD S.T.E.A.M. WITH GREAT BOOKS

SCIENCE

So That's How the Moon Changes Shape
By Allan Fowler

Man on the Moon
By Anastasia Suen

TECHNOLOGY

Cars on Mars: Roving the Red Planet
By Alexandra Siyon

Looking Through aTelescope
By Linda Bullock

ENGINEERING

Henry Ford: Big Wheel in the Auto Industry
By Mike Venezia

What Do Wheels Do All Day?
By April Jones Prince

ART & MUSIC

Art & Lots of Dots
By Ginett Alarcon

Eight Hands Round: APatchworkAlphabet
By Ann Whitford Paul

MATH

When a Line Bends . . . A Shape Begins
By Rhonda Gowler Greene

Shapes in Math, Science and Nature: Squares, Triangles and Circles
By Bill Slavin

ELA

Ladder to the Moon
By Maya Soetoro-Ng

Papa, Please Get The Moon for Me
By Eric Carle

SCIENCE

CIRCLE TIME
STEMVESTIGATION: WEATHER AND WIND

READ **Three Mousy Sisters**

DISCUSS What happened to the little mice? What could they have done to prevent problems during the rainy season? What would you do to prevent the destruction?

EXPLAIN Weather is the day-to-day conditions, especially temperature, cloudiness, and rainfall, affecting a specific place. Weather changes from hour to hour or day to day. It is affected by temperature, humidity, gravity, wind, and many other factors.

MOVE Play the thunderstorm game. Sit cross-legged on the floor. Half the group will be rain. The other half will be thunder. Begin telling the story of a light rainfall moving in as it pitter-pats. Students lightly slap hands on knees. The rain gets harder. Knee slapping gets harder. Thunder rolls. The other half of the class claps their hands 3 times. See activity page instructions.

INVESTIGATE

Begin collecting data on wind speed, wind direction, temperature, and rainfall. Continue filling the data tables through the school year.

ELA Make a classroom library of favorite books featuring wind, rain, snow, and other meteorological phenomenon.

STEMVESTIGATION: WEATHER

Draw the weather each day for the week.

Three Mousy Sisters

Explore structures in science

Three Mousy Sisters

Explore structures in science

This is the story of three little spiny mice. One summer day when heat rolled down the rooftops to crash on dusty dry ground, Baba Mouse said to his three little darlings, "Clever daughters, it is time to make your way in the world. Build strong houses. Rain is coming." The three spiny mouse sisters set off toward Baba's land on the mountainside. Oldest Sister was the first to tire of walking. She found a fl at area near the dry, dusty riverbed full of sticks. She wove the sticks together like a basket. She was lazy, so she dropped for a nap before she finished the stick house completely.

Halfway to the top of the hill, Middle Sister tired of walking. The day was hot and sunny, and she wanted to decorate her house with silk scarves from the market. She also made a stick house but had enough sense to make a wattle and daub house by filling in all the holes with mud. Before the mud daub dried, Middle Sister wandered away to pick flowers in the sun. Little Sister walked all the way to the top of the mountain. She spent all day toiling in the hot sun. She sketched a design in the dirt and made a little mud model before beginning work on her real house. The model's roof fell, so she made a few changes in her plan. She formed clay walls, adding layer on layer until her house was finished. Its walls were strong. The re-designed roof was tight and no leak of rain would reach her tiny mouse bed of grass.

The sisters settled happily into their beds that night. At five a.m. the rains came. Water dropped from the sky to pound rooftops and make the very ground shake like a drum.

Oldest Sister heard a rush of water in the dry riverbed where she'd built her house. The ground shook, but Oldest Sister escaped to scramble up the bank and run for up the mountain. Water rushed down the once-dry riverbed, taking oldest sister's house with it.

Middle Sister's house was dry. She watched the rain sluice the hill. Oldest daughter ran into the house. Rain fell hour after hour until the ground was soaked. Mud began to slide down the hill. Crash! Crash! The sisters scrambled out of the house as mud knocked down the walls. The spiny mouse sisters slipped and slid their way up the hill to the home of Little Sister.

Little Sister opened the sturdy door and handed both sisters dry blankets made of rushes. She helped them find comfortable places to sleep and they waited for the sun to come again. Weeks passed as they spun wool, and read books, and ate jars of preserved food. The rain stopped as suddenly as it had begun. The mousy sisters helped each other build houses side by side at the top of the hill. That is how the mouse learned to work hard while the sun shines.

TECHOLOGY

CIRCLE TIME
STEMVESTIGATION: PATTERNS

READD The Shepherd's Crook

DISCUSS Hickory, Dickory, Dock is a song with a meaning. The first people to use the rhyme were shepherds. They took care of what kind of animal? How did they use thumbs to count?

EXPLAIN Hickory, Dickory, Dock is like a lot of other nursery rhymes in that it started out as a grown-up story. Because a lot of people in the 1700's could not read or write, the farmers had to find other ways to trade food, animals, and other things. They were creative in finding ways to "tally" numbers.

MOVE Sing Hickory, Dickory, Dock, and act out counting on thumbs and crook.

INVESTIGATE

Create a tool that helps you count to 10. Would you make a shepherd's crook like the nursery rhyme? Would you use a bag of pebbles? Would you make marks on a board? How could you keep track of counting so that neither you nor your partner can make any mistake.

ELA Learn the words to Hickory, Dickory, Dock.

Count to ten. Circle 10 buildings.

Draw 10 objects.

Hickere, Dickere, Dock or The Shepherd's Crook

Hickere, Dickere, Dock

Explore counting.

Hickere, Dickere, Dock was first published in 1777 in Ireland. Hickere, Dickere, Dock meant 8, 9, and 10 in the Westmoreland dialect of England.

Shepherds in those days used a special system of counting sheep. They could not count past 5, so their system broke numbers down into small groups of 5.

Touch each knuckle on right hand with a thumb. That's 1-2-3-4-5.
Hand moves up a notch on the crook.

Touch each knuckle on left hand with a thumb. That's 1-2-3-4-5.
Hand moves up a notch on the crook.

Touch each knuckle on right hand with a thumb. That's 1-2-3-4-5.
Hand moves up a notch on the crook.

Touch each knuckle on left hand with a thumb. That's 1-2-3-4-5.
Hand moves up a notch on the crook.

When the shepherd reaches the fourth notch, he has counted to twenty. He drops one stone in a counting bag (the clock strikes one) and drops his hand down to begin counting again.

NURSERY RHYME	HIDDEN MEANING
Hickory, Dickory, Dock	8, 9, 10
The mouse ran up the clock.	The thumb counts up the shepherd's crook.
The clock struck one.	The count reaches 20, and one stone is dropped in the bag.
Away he run.	The hand moves back down the shepherd's crook to start again.
Hickory, Dickory, Dock	8, 9, 10

Try counting like the Irish shepherds on a crook of your own.

ENGINEERING

CIRCLE TIME
STEMVESTIGATION: WHEELS

READ **Wheels on the Ground**

DISCUSS What is a wheel? Wheels rotate around the axle on a vehicle. Look at a wagon, roller skate, skateboard, bicycle, classroom cart, and other rolling objects. Rolling chairs are trickier and but make great examples. The axle is often hidden under a plastic cover.

EXPLAIN Wheels are connected to an object on a rod called the axle like those on bikes, cars, or a set of roller-blades. Wheels are also used as doorknobs, clock gears, on pulleys, and in lots of mechanical appliances.

MOVE Drag a student seated in a box across the floor. Now pull students in the rolling chair. Do a relay. One line of students will pull the box without wheels. One line of students pulls the box on wheels (the rolling chair). Time, graph (x=time, y=distance), and compare outcomes.

INVESTIGATE

Make pinwheels! Given a pile of potential fabrics, papers, foils, cardboards, and other items, let students choose the material for their pinwheel. Use the handout template as a tracing pattern.

Challenge: Design and create a pinwheel that turns at least 20 times in one minute (use a sand timer, stopwatch, or the small hand on the big clock).

Create a chart that compares each material to its outcome. example

material	turns in 1 minute
cardstock	20

ELA Write and illustrate shape poems inside circles.

STEMVESTIGATION: PINWHEELS

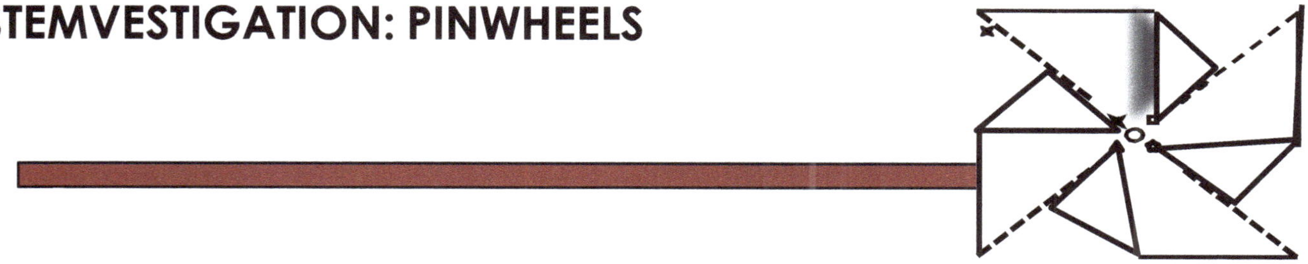

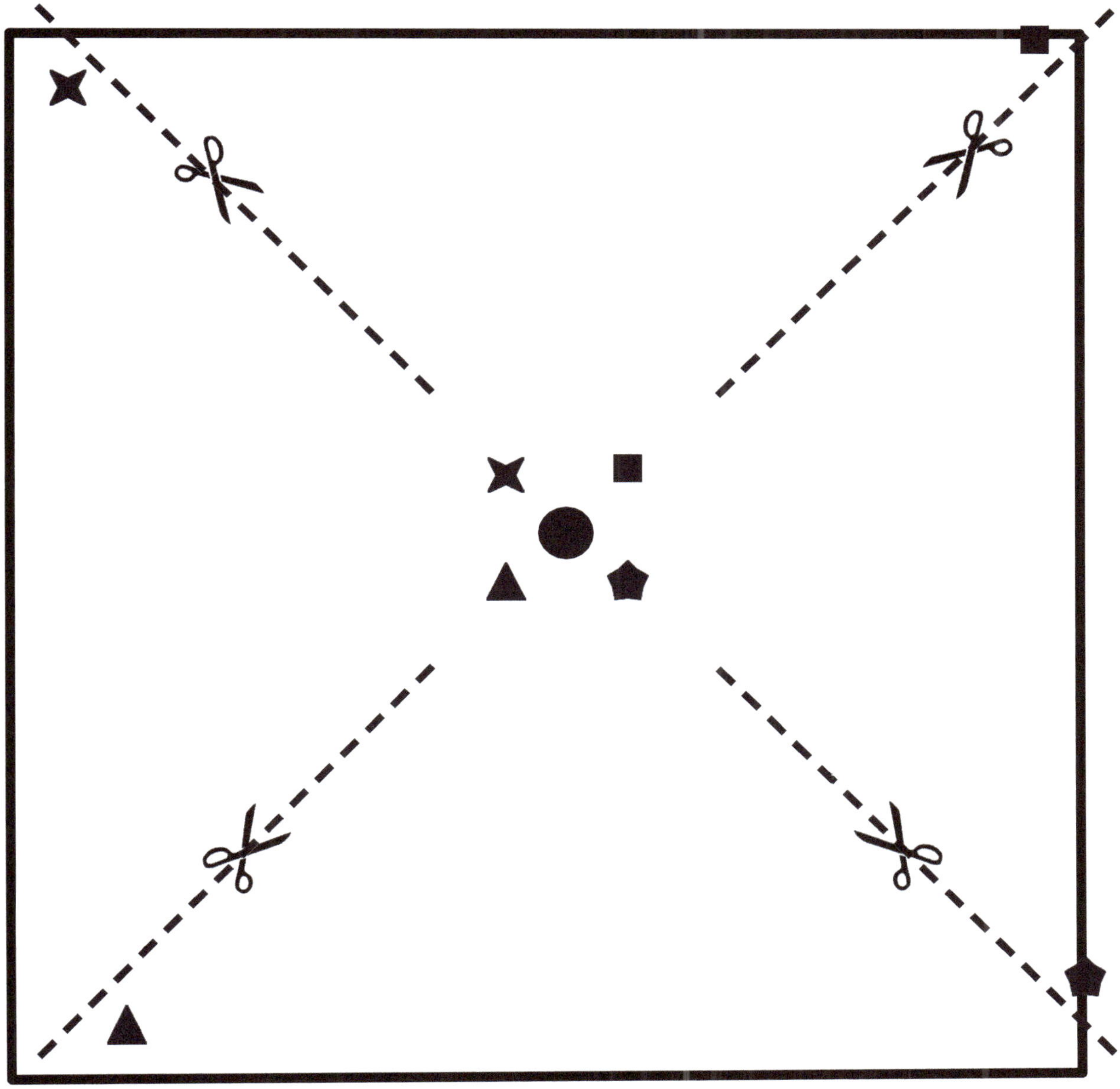

Cut on the dotted lines.

Fold shape to like shape. ▲⟶▲

Use a glue dot or tape to attach each shape to like shape.

Wheels on the Ground

Explore circles.

Wheels on the Ground

Explore circles.

Radius and Pi were friends. They looked the same. They even dressed the same, but Radius and Pi were very different people.

Radius loved things that were fast!

He did not stop to think about what could happen. Pi was a planner. She made lists and always made decisions after lots of thinking. Radius and Pi made a good team.

On Saturday, the twins always pulled a cart of organic fresh vegetables and fruits to the local Farmer's Market. The twins grew their own food to sell. They set up a table beside all the other farmers and gardeners. Radius and Pi never missed a market. This weekend Radius and Pi had a problem. The rain had made muddy puddles on the market path. The cart's skinny tires had sunk 2 inches in the mud. The twins were stuck.

Pi sat down to think. She made notes on a pad and listed all the ways they could solve this problem. She came up with trucks and sleds.

Radius dove into his project. He liked the idea of a sled. He found a big piece of plastic and piled all the fruits and veggies on the sled. Radius yelled, "Ta Da!" He pointed to his solution and dragged the sled a few feet. It worked but was heavy.

Pi decided to try the truck idea. She found four tires from a toy in the shed. Ra-dius helped Pi prop up the cart. They used a screwdriver and wrench to remove the cart's wheels. Radius held the cart steady as Pi connected each wheel to the cart's axle. The twins lowered the cart to the ground.

Radius gently pulled the handle. He tugged a bit harder. Finally, the cart started moving. The new, wider wheels worked pretty well! Their cart sank only about 1 inch into the mud. The twins moved all the produce to the wagon. The wagon rolled along the path all the way to market.

Radius and Pi sold all the food and had a great day. Best of all, they had worked as a team to solve a tricky problem. Radius and Pi always made a good team.

ART

CIRCLE TIME
STEMVESTIGATION: EXPLORE PATTERN QUILTS

READD The Circle Tree

DISCUSS How did the author introduce the tree? Why did the tree lose its leaves? What did the animals do to make the tree feel joy again?

EXPLAIN Circles measure the same distance from center to outer edge all the way around. If you fold a circle down the center, both sides line up at the outer edges. That means that circles are always symmetrical. You can rotate or flip a circle, and the circle keeps its same shape.

MOVE Form a classroom circle. Have two students raise their joined hands to make a bridge. The whole class keeps hands joined as the pairs of kids pass under the bridge. When the whole class has passed under the bridge, your circle will have turned inside out... Students will face out rather than in toward the center.

INVESTIGATE

Incorporate your current class topics in creating a Topic Tree that resembles the one in "The Circle Tree." Students use circle patterns to cut their own circles, or use two sizes of paper plate. The center circle should represent your theme. Use Velcro dots to affix the topic circles, and students can take their pieces home at the end of the month. Start a new Topic Tree for squares or triangles, or keep the circles and incorporate a new reading / math / science / history / holiday theme each month. These also make great word walls.

ELA Read the story aloud in groups.

RHOMBI'S CIRCLE TIME
STEMVESTIGATION: CIRCLE TREE

Draw the circle you will make.

Cut out these circles to use in your tree.

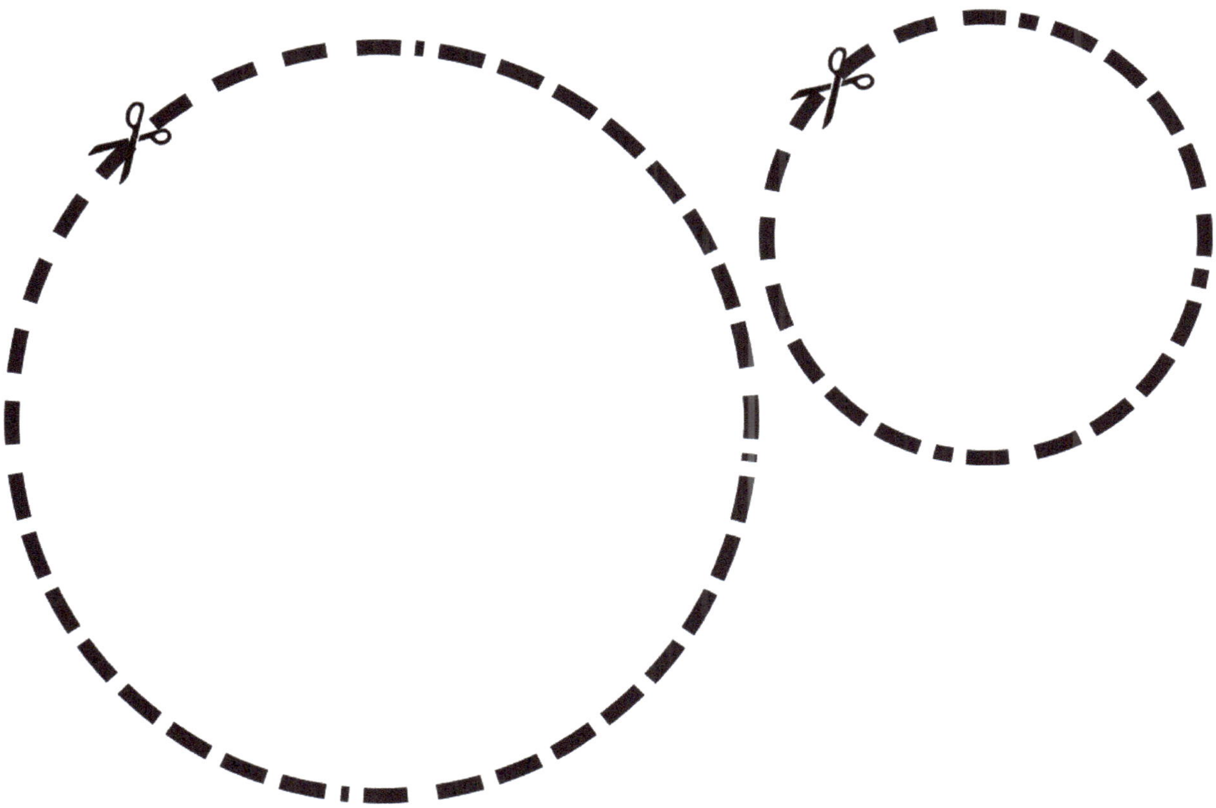

The Circle Tree

Explore circles.

The Circle Tree

Explore circles.

There once lived a tree in a forest up the hill. It welcomed any animal that Needed a home. Its full branches held squirrels and birds and bees- and bugs inside the lush green leaves. In Fall, the tree decorated the forest with its leaves of gold and red. When Winter came, the tree tucked in its branches to shelter squirrels and raccoons from snow and ice until the spring thaw.

The tree returned each Spring with a burst of leafy green branches that fed cat-erpillars and hid baby birds until they were ready to fly. Every animal in the forest had been helped by the tree in its 200 years. Finally, though, the tree's branches began to thin. Its bark grew brittle and broke off in great flakes. When Winter winds blew, the tree shivered. When Spring arrived, there were no leafy green hiding places for tiny animals.

It was no longer home to dozens of animals, but the tree still stood tall and proud. One morning it heard singing. The tree swayed to music that played far down the hill at an end-of-the-school-year art party. Kids painted thick round paper with bright colors. They added paper flowers and ribbons or yarn. Each child wrote the happiest word they could imagine. It was a cheerful tradition that ended every school year.

The school picnic was about to start when gust of wind snatched the school's art projects. It sent them spinning off into the forest. Laughing children and their teachers ran after the round plates. They chased the beautiful art all the way to the foot of the ancient tree. The Principal pointed at the tree. On every branch hung a piece of art. The whole school sat under the tree to eat their picnic. Final-ly, all the kids headed back down the hill toward waiting buses. The forest was calm again. Singing. Cheeping. Chirping. Chittering. The air around the tree was filled with happy sounds. Birds grabbed yarn for their nests. Squirrels snagged papers to line their homes.

A new tradition started. Every year, the school ended their year by hanging good thoughts on bright round art in the tree's branches. The tree was very happy.

MATHEMATICS

CIRCLE TIME
STEMVESTIGATION: EXPLORE RADIUS AND CIRCLES

READ **Radius and Pi**

DISCUSS Make a list of round shapes. Put them in order from smallest distance from center to outer edge. Radius and Pi need a wheel that fits their market cart. Will a skate or chair wheel fit?

EXPLAIN Circles are included in the oval category. All points on the bent line are the same distance from the center of the circle. If you hold a string in the center of the circle, you should be able to pull the string out to any place on the circle. If the string is too long or too short at any place, the shape is an oval... not a circle.

MOVE Use string to check for circles or ovals in the classroom or playground.

INVESTIGATE

Measure the radius from the center of your head to the tip of your middle finger. (Measure from fingertip to fingertip to find the diameter of "your circle.") Cut a length of string measured to the length of "your radius." Pin "your radius" string to the middle of a large sheet of paper. With a pencil at the other end of the string, drag the string around in a complete circle.

ELA Write and illustrate shape poems inside the circles.

CIRCLE TIME
STEMVESTIGATION: RADIUS

Draw the shape you made with your string.

Radius and Pi

Explore circles.

Radius and Pi

Explore circles.

Radius and Pi rolled along the path toward Saturday market. Suddenly, the cart stopped. Radius stopped and circled the cart looking for a problem. He leaned down to look at the wheel and axle. The wheel had wobbled loose and cracked. Radius looked around for some way to fix the cart. The twins needed a new wheel. Radius and Pi began looking for round objects that could fit the cart.

The twins found a roller skate wheel and a skateboard wheel in a pile by the curb. Both wheels would be too small. The length from the center hole to the outer part was just too short. Their cart might roll, but it would tilt and spill the apples.

They found some balls, but none had a hole in the center. No hole meant no axle.

How would the ball stay on without an axle down the middle?

Radius tried to hold up the no wheel corner while Pi steered. The two entrepreneurs really wanted to reach the market in time to sell their fruit. They moved too slowly without a wheel. Pi and Radius looked at the bicycles in a shop window. They looked at each other with wide grins and carefully propped the cart on a step.

Four minutes later, Radius emerged with a bicycle wheel minus the tube. It was old and rusted but not bent. The shop owner gave them the wheel. The twins gave her ten apples. The wheel was just the right length from center to rim.

Radius and Pi put the wheel on their cart. They zoomed to the market. When the market gate opened, customers were happy to see the twins. Radius and Pi sold all the yummy ripe apples. The twins were proud that they'd solved an engineer-ing problem all on their own.

THE BIG WIND BLOWS

MATERIALS
none

SETUP
One person is chosen to be the "wind." Wind stands in the center of the circle.

INSTRUCT THE PLAYERS
The game begins when the person in the middle begins turning in a circle and waving arms. The wind says, "THE BIG WIND BLOWS (insert a statement that is true about himself or herself."
ie "The Big Wind Blows everyone who has red shoes."

All of the kids who have red shoes including the wind stand up and scoot through the circle to a position that is now empty on the other side. They sit down in the new spot. The one person left over is now the "wind." Play continues.

There is no winner or loser, just a lot of fun.

UNIT 1
CIRCLE TIME

Once upon a time, there was a kid named Rhombi. Rhombi liked shapes. She saw shapes all over her home and school.

Rhombi's favorite shape was the circle. Rhombi had a list of all the circles she saw. A lot of things on the list could spin. Skateboards, bicycles, and the giant Ferris wheels at the county fair were all circles. They could spin.

Rhombi said out loud, "All the things on my list that roll and spin are round. I wonder if all things that spin are round?" Rhombi was walking to Uncle Reel's bike shop. Her bicycle's round wheel was bent, and Uncle was fixing it for her.

When Rombi walked into the store, her uncle had turned her bike upside down. He pulled the tire to make it spin. Rhombi asked, “Uncle Reel, what are you doing?”

"Hello, Rhombi. The wheel needs to spin evenly so your bike is easy to steer," Uncle Reel said with a smile. "Here you go!"

Rhombi hugged him and said, "Thank you so much! I owe you a favor."

Rhombi wanted to give Uncle Reel a gift to say thank you for fixing her bike. She knew she would need help with the project. She looked out from the pages of her world and asked, “Designers, will you help me create a round spinning toy for Uncle Reel?”

RHOMBI'S ADVENTURES IN 2D
TALK ABOUT S.T.E.A.M.

Name some things on Rhombi's list that have wheels.
Skateboards, Ferris wheels, bicycles

How does Rhombi get to her uncle's shop?
She rides a bicycle.

What is Rhombi's uncle's name?
Uncle Reel

What type of shop does Uncle Reel own?
bike shop

Why was Rhombi going to Uncle Reel's shop?
Her bike wheel was bent. She needed to pick up her bicycle.

What is another word for rotation?
spin

What does Rhombi want to make?
She wants to give her uncle a gift that rotates / spins...

What is a designer?
Someone that comes up with ideas and finds ways to make them work...

"Designers, will you create a toy that rotates? My uncle likes things that spin."

my idea	my plan

RHOMBI'S ADVENTURES IN 2D
UNIT 1 RUBRIC: CIRCLE TIME

TASK **Students will use their knowledge of circles to design, troubleshoot and build a spinning or rotating toy.**

	Content	Organization	Presentation
1	• Is well thought out and supports the solution to the challenge or question • Reflects application of critical thinking • Has clear goal that is related to the topic • Is accurate	• Information is clearly focused in an organized and thoughtful manner • Information is constructed in a logical pattern to support the solution	• Student is able to accurately discuss what makes the toy spin. • Student shows the toy and explains why its design was chosen.
2	• Supports the solution • Has application of critical thinking that is apparent • Has no clear goal • Has some factual errors or inconsistencies (i.e. wheels do not rotate around axle)	• Project has a focus but might stray from it at times (more concerned with form than function) • Information appears to have a pattern, but the pattern is not consistently carried out in the project	• Student says the toy spins but does not say why. • Student does display but does not attempt to explain reasoning behind the toy.
3	• Provides inconsistent information for solution • Has no apparent application of critical thinking • Has no clear goal • Has significant factual errors, misconceptions, or misinterpretations	• Content is unfocused and haphazard • Information does not support the solution to the challenge or question • Information has no apparent pattern	• Student is unable to talk about the toy. • Student forgets to show the toy.

Teacher's notes ABOUT THE CIRCLE

- In the math world, circle refers to the boundary of the shape. "Disk" is used to refer to the whole shape, including the inside.
- A straight line from the center of a circle to the edge is called the radius.
- A straight line that passes from one side of a circle to the other through the center is called the diameter.
- The distance around the outside of a circle is called the circumference. All points on the edge of a circle are the same distance to the center.
- The value of Pi (π) to 2 decimal places is 3.14, it comes in handy when working out the circumference and area of a circle.
- Circles have a high level of symmetry.

UNIT 2
SQUARE ONE

RHOMBI'S ADVENTURES IN 2D SQUARE ONE

Essential Questions

1. What are the attributes of a regular square?
2. How does technology help you learn about new things?
3. What strategies make counting easier?
4. When would a pattern make a task simpler?
5. What are you trying to accomplish in Rhombi's Challenge?

ENDURING UNDERSTANDING

Two dimensional shapes create new shapes.

"Halves" means two equal parts.

Circles and rectangles break into equal parts.

Breaking circles or rectangles into more equal parts means that the parts will be smaller.

Three objects may be put in order from longest to shortest by comparing their lengths.

The length of an object may be discussed using whole numbers.

A smaller object may be used as a measurement tool when measuring longer lengths.

Larger numbers consist of tens and ones.

Organizing information helps in discussion.

SQUARE ONE

Focus on polygons. Use tallies to "harvest" shapes at lunch. Employ technology in discovering discussing geometry. Sharpen observational skills with shape walking. Research and recreate famous quilted art. Make your own origami square papers, and learn to fold boxes.

Students will apply basic measurement, understanding of number, place value, data collection, and data application in problem solving.

Rhombi needs to repair the roof of her toy barn. Students design a square roof that will support 1/2 pound of weight.

SQUARE ONE
CURRICULUM CONNECTIONS

Rhombi needs to repair her toy barn's roof.

MATH BACKGROUND - Squares

Squares are -by definition- regular polygons in that all edges must be the same length. Because they have 4 sides of equal length, squares are regular quadrilaterals. Squares are also rectangles with equal sides and rhombuses with right angles. Square shapes are often used for design and engineering purposes such as town planning.

TECHNOLOGY CONNECTIONS

Build logic and spatial reasoning.

- GOOGLE: Math Playground
- GOOGLE: Math Fun

Mr. R's Math Videos and Songs

- GOOGLE: R Math Songs

The process of quilting uses a needle and thread to join two or more layers of material. During pioneer days, "paper" quilting became popular. Each piece of cut fabric was basted around paper patterns. Paper was scarce in the west, and women would save letters from home, newspaper clippings, and catalogs to use as patterns. The paper acted as pattern and insulator. Paper found between old quilts is a primary source of information about pioneer life. During the Civil War, slaves used quilts as a means to share and transmit secret messages.

ART CONNECTIONS

Piet Mondrian was born in Amersfoort in the Netherlands, in 1872. Mondrian began his career with traditional art styles, but embraced the concept of "cubism." His most famous painting was "Composition with Red, Yellow and Blue" painted in Paris in 1921.

Cubism was developed by Pablo Picasso and George Braque in the beginning of the 20th century. A typical Cubist painting depicts real people, places or objects, but not from a fixed viewpoint. Instead it will show you many parts of the subject at one time, viewed from different angles, and reconstructed into a composition of planes, forms and colors.

READING CONNECTIONS

Introduce each activity and lesson with a book. Books are listed on the Read about STEAM page. Jot additional lists of books from your favorites here.

1.

2.

3.

4.

5.

SQUARE ONE
CURRICULUM CONNECTIONS

COMMON CORE CONNECTIONS

ELA/Literacy

W.K.7 Participate in shared research and writing projects (e.g., explore books by a favorite author and express opinions). (K-ESS2-1) add Venn

Mathematics

MP.2 Reason abstractly and quantitatively. (K-ESS2-1)
MP.4 Model with mathematics. (K-ESS2-1)
K.CC.A Know number names and the count sequence. (K-ESS2-1)
K.MD.A.1Describe measurable attributes of objects, such as length or weight. Describe several measurable attributes of a single object. (K-ESS2-1)
K.MD.B.3 Classify objects into given categories; count the number of objects in each category and sort the categories by count. (K-ESS2-1)

NEXT GEN SCI CONNECTIONS

ETS1.A: Defining Engineering Problems
A situation that people want to change or create can be approached as a problem to be solved through engineering. Such problems may have many acceptable solutions. (Secondary to K-PS2-

1-ESS1-1,1-ESS1-2 Patterns in the natural world can be observed, used to describe phenomena, and used as evidence.

1-ESS1-1, 1-ESS1-2 Make observations (firsthand or from media) to collect data that can be used to make comparisons. Analyzing data in K-2 builds on prior experiences and progresses to collecting, recording, and sharing observations. Use observations (firsthand or from media) to describe patterns in the natural world in order to answer scientific questions.

UNIT 2 S.T.E.A.M. ACTIVITIES FOR "SQUARE ONE"

Rhombi Audio Download / Video (available December 2014)

UNIT2: Square One: Science:	Look for food shapes.
UNIT2: Square One: Technology:	Investigate the world with magnifying tools.
UNIT2: Square One: Engineering:	Take a shape walk to look for squares.
UNIT2: Square One: Art & Music:	Discover the geometry of quilts.
UNIT2: Square One: Mathematics:	Explore paper squares, and fold origami boxes.
UNIT2: Square One: ELA:	Read "Square One." Students illustrate the story.

ON THE SQUARE
BUILD S.T.E.A.M. WITH GREAT BOOKS

SCIENCE

Pick a Circle, Gather Squares: A Fall Harvest of Shapes
By Felicia Sanzari Senesky

SQUARES (Shapes in Math and Nature)
By Sheldrick Ross, Catherine and Bill Slaven

TECHNOLOGY

The Kids' Guide to Digital Photography: How to Shoot, Save, Play with & Print Your Digital Photos
by Jenni Bidner

Looking Through aTelescope
By Linda Bullock

ENGINEERING

Finding Shapes with Sebastian Pig and Friends at the Museum
By Jill Anderson

Shapes : squares
By Sarfatti, Esther

Shapes at School
By Joyce Jeffries

ART & MUSIC

Art & Lots of Dots
By Ginett Alarcon

Dare to Be Square Quilting: A Block-by-Block Guide to Making Patchwork and Quilts
By Boo Davis

MATH

Sam's Sneaker Squares
By Math Matters & Nat Gabriel

My Mother Taught Me Origami
By Manorama Jafa

ELA

Square Bear Meets Round Hound
By Marianne W. Penn

Doodle Dandies: Poems That Take Shape
By J. Patrick Lewis

SCIENCE

SQUARE ONE
STEMVESTIGATIONS: GROWING THINGS

READD

READ	**Square Foot Harvest**
DISCUSS	What do students already think they know about harvesting? Make a list to identify "mindbugs."
EXPLAIN	When a crop has grown and is ready to eat, the farmer must collect it from the fields. This is known as the harvest. Grain crops are collected by huge combine harvesters. Some tree fruits are picked by tall machines. Various crops are harvested all year.
MOVE	Pretend to harvest various foods. Will you stretch tall to reach an apple? Should you bend deep to pull carrots or peanuts? Call out a food, and students move through the motions.

INVESTIGATE

"Harvest" the shapes you can find at lunch. What vegetables and fruits are present in student lunch boxes or on their cafeteria trays? Work with students to create a chart for tallying the various foods. Students should offer their educated guesses as to which foods will be represented. After tallying lunch items, compare the guesses with the actual tallies. This is an opportunity to combine counting, data representation, data analysis, geometry and data collection in a simple activity.

ELA	Imagine the harvest using five senses. Describe colors, shapes, tastes, sounds, and textures. Students close their eyes and list a descriptive word or phrase per sense.

Tally the vegetables and fruits in your lunch.

Day	Fruits	Vegetables	Dairy	Protein
Monday				
Tuesday				
Wednesday				
Thursday				
Friday				

Square Foot Harvest

Explore garden harvests in science

Square Foot Harvest

Explore garden harvests in science

Leaves crunched underfoot as Rhombi headed toward the garden. Rhombi and her friends had planted vegetable seeds in the Community Garden in August. Over the past two months, the seeds pushed out of their hard seed covers. In September, the seedlings climbed through dirt toward the sun.

Rhombi had spent many hours making sure that each seedling had sun and room to grow. When she saw leaves, Rhombi knew that roots had reached down into the soil. Finally, after almost days, the radishes were ready to eat.

Carrots and radishes and lettuce waited for Rhombi. She pulled 2 carrots and put them into her basket. Next, she layered 2 radishes. Finally, Rhombi added four heads of lettuce. She picked up her heavy basket and headed home through the crunchy leaves.

She washed and chopped carrots and radishes. She tore lettuce leaves and added them to a big bowl. When her friends sat down to dinner, Rhombi would proudly serve her delicious autumn harvest salad!

TECHOLOGY

SQUARE ONE
STEMVESTIGATIONS: WATCH AND LEARN

READ **Fox on a Log**

DISCUSS Fallen trees become home to many insects, plants, fungi, and small animals. The forest floor supports thousands of species.

EXPLAIN Forests are unique to each part of the world, but they all have a few things in common. View the virtual forest tour for a glimpse into biomes, habitat, forest creatures, and forestry jobs.

MOVE Play "Going on a Quad Hunt." In the spirit of "Going on a Bear Hunt," sit in a circle with legs folded. Slap legs in a rhythm (slap, slap, clap, clap, slap, slap, clap, clap). Each person in the circle adds a scene to the chant. For a sample, visit Schooltube for videos.
http://bit.ly/rtpbNU or
www.sophie-world.com/activities/singalong-were-going-bear-hunt

INVESTIGATE

Use magnifying glasses and binoculars to watch what's happening outside the classroom window. This is a perfect time to set up a long-term station rather than a whole-class exercise. Encourage students to keep a class log full of pictures and words to describe what they observe.

After a week, students should predict the animals they might see using the "data" they've collected in the class logbook.

ELA Present findings to the class.

SQUARE ONE
STEMVESTIGATION: MAGNIFIER TOOLS

Draw the objects you find in the habitat.

1.

2.

3.

Fox's Forest

Explore environments in art

Fox's Forest

Explore environments in art

Fox trotted happily on the warm summer path. Sun beamed down through leaves to make moving patterns on the worn dirt track. Fox's head swung from side to side looking for yummy treats. He also scanned the woods for anything that might want to eat him!

Bushy tail wagging, Fox stopped to nose some flowers. Soon these green leaves would part to reveal little red berries. Delicious and sweet, Fox loved these berries in the straw. Fox caught another smell in the straw. It was the ripe odor of rotting log. Logs were good. They often held tasty, crunchy bugs.

Fox lifted his little pointed face to the wind. Here was the smelly log He faced west toward the setting sun. Sniff, sniff. Fox smelled logs there. He faced east toward the far edge of the woods. Log smell rose from that direction. Fox turned north to sniff. There. The stinky log was that way!

Nose to the ground, Fox left the path and headed into the green woods. Shadows played across o's rust colored body. Sniff, sniff. He followed the air molecules that blew in the breeze from that stinky log. The molecules were pulled in through Fox 's nose and mouth. His brain made sense of the smells, and Fox turned a little to the east. Eventually, behind a tall rock, Fox found a log so rotten it made his mouth water. Using tiny paws and claws, Fox pulled at the wet bark. Beetles and mealworms tried to scatter. Fox caught them and crunched his way through an afternoon snack. Delicious!

With his pointy little face covered in soil and bits of rotten tree, Fox turned back toward the path. South and a little East brought Fox right back to straw-berry leaves beside the warm summer path. It wasn't quite so warm now that the Earth was turning away from the Sun. Shadows crept closer as Fox found his way back to the den. With a full belly, Fox curled up with his tail over his nose and fell asleep in the safety of his dark, dry hole.

ENGINEERING

SQUARE ONE
STEMVESTIGATIONS: SHAPES WALK

READ **Shopping for Shapes**

DISCUSS From chairs to skyscrapers, every object has a shape. Often, several shapes are hidden in plain sight. Brainstorm a list of locations at which kids see shapes every day.

EXPLAIN Squares have uses in architecture, engineering, and technology. The square must have edges of equal length and 4 right angle corners (vertices).

MOVE Use bodies, arms, legs and books to make shapes as they are called out by "Rhombi". Rhombi says, "Show me a square." Rhombi says, "Make your arms into a circle."

INVESTIGATE

What squares can be found in your school? Create a comprehensive list of square objects and objects that incorporate squares. How will you check for square shape vs. not? Not all things that appear identical are really the same. Hint: remember to look up and down when square hunting. Take a walk to shop for shapes!

ELA Make a diorama of your school or a classroom using cut shapes and glue sticks!

Take a shape walk. Look for squares.
Draw or describe.

Look up!

Look down!

Look around!

Shopping for Shapes

Explore shapes in engineering

Shopping for Shapes

Explore shapes in engineering

Poly and her friend Trap were shopping for groceries. The two were meeting up with Rhombi that afternoon for chess club. It was Poly's turn to bring snacks. She had ten dollars to spend.

Poly pushed a cart up and down the aisles looking for something her friends might like to eat. She dropped a pack of gum in the cart, but it fell through the squares that covered the cart. She placed the gum on the rectangular face of a box of cookies.

Poly wanted to find cheese, crackers, fruit, and cookies. So far, she'd found cookies.

They were the round kind with chocolate chips. Trap held up a bag of $3.00 star fruits. They looked like spheres or balls now. When cut, the star fruits would look like little green star shapes on a plate. He also had fat purple grapes that cost $1.00 per bag.

The list included cheese, crackers, fruit and cookies. So far, Poly and Trap had cookies and fruit. The two friends headed toward the dairy section to look for cheese. Wow, there were a lot of cheese shapes. Which one should they choose? There were cylinders of string cheese. Poly pointed to brick shaped rectangular prisms of cheddar. Trap spied a bag of already cut cubes that would be easy to pile on a tray. The $2.00 cheese cubes were added to the cart.

The last item on their snack list was crackers. The cracker aisle was filled with boxes of all shapes and sizes. Oval buttery crackers sat next to 2 inch square saltines and little ½ inch round soup crackers. There were whole wheat rectangles woven like blankets and cheesy 1 inch squares. Some were even shaped like goldfish. Trap picked the $2.50 woven rectangles. They tasted good with cheese.

Their cart was filled with delicious shapes. Star fruits, grapes like spheres, cubes of cheese, and square crackers filled the cart. The total cost was $8.50. They had one dollar and 50 cents left. Poly and Trap took their bags and headed to the chess club. Snack time shaped up to be a yummy break.

SQUARE ONE
STEMVESTIGATIONS: QUILT PATTERNS

READ **Quilting in Circles**

DISCUSS Quilts are generally geometric designs created using fabric shapes. Quilt patterns may be simple repeating squares or circles. Patterns may also be complicated.

EXPLAIN Quilts use all shapes as part of images sewn to add interest. Modern quilts have much brighter fabrics than older quilts and often boast more boldly geometric patterns.

MOVE Bring a quilt (or blanket) to play. Students stand around the edge of the square quilt holding its edges with both hands. Pick up the quilt. The teacher drops a balloon on the quilt, and students must work together to pop their balloon up into the air and keep it on the quilt.

INVESTIGATE

Using patterns from a quilting source, use paper to cut-and-paste quilts instead of sewing them. Recreate famous quilt patterns, research modern patterns, and use the resulting geometric artwork as a way to discuss patterns, shapes, measurement, and organization.

Read the story aloud in groups.

ELA Write fictional stories about the paper quilt owners.

ART

STEMVESTIGATION: QUILT PATTERNS

Use triangles, circles, and square to draw a pattern.

Quilting in Circles

Explore circles

Quilting in Circles

Explore circles

Mole and Mouse are sewing a quilt. They planto stack three layers, one on top of another. The friends start with a bottom layer using squares of old t-shirt fabric. The center layer is fluffy cotton from pillows. The top layer is bright fabric with circles.

Each friend has a thimble to keep fingers safe from sharp needles. They use a pattern to cut many circles in four different sizes. Mole cuts circles that measure 1 inch from the center, 2 inches from the center, 3 inches from the center, and 4 inches from the center.

Mole draws a pattern of circles that look like a rainbow of dots on the fabric. Some of the circles touch edge to edge. Other circles sit one inside another. Quil-ters call this pattern an echo. Each circle is sewn by hand. Each circle is added one by one.

The planet is rotating away from the Sun when Mole and Mouse finish the proj-ect. The Sun sinks out of sight, and they see the round full moon in the sky. The twins put down their needles. They shake out the colorful quilt. Circles overlap and dance on one side. Bright t-shirt squares fill the other side.

On the day of their library fundraiser, Mole and Mouse sell raffle tickets. Friends and family donate a dollar and get a raffle ticket in return. At the end of the day, someone will win the quilt made by Mole and Mouse.

Mole and Mouse sell 100 raffle tickets and make $100 for the library. At the end of the day, Mole pulls one of the tickets out of the bowl. Mole starts to giggle. Fi-nally, she calls out the winner's name. It's Mouse! Mouse has won her own quilt.

MATHEMATICS

SQUARE ONE
STEMVESTIGATIONS: ORIGAMI

READ **Paper Forests**

DISCUSS What is origami? Origami is the art of folding paper to look like specific shapes, animals, boxes, and other items.

EXPLAIN Origami papers are shaped like squares, because that shape offers the widest variety of symmetrical folds. Both halves of any origami shape are identical. If the origami bird has a left wing, it will also have a right wing.

MOVE Fold a square! Fold a triangle! Begin with any piece of notebook or construction paper. Fold the top corner down so that A and AA line up exactly. Cut along the resulting line to remove excess paper. Unfold the triangle to reveal a regular square.

INVESTIGATE

Make your own origami square papers. What shapes can you fold from the starting point of a simple square? Experiment A with various paper thicknesses. What happens when you fold the same shape from three different types of paper (construction, lined, tissue)? Students should journal about each paper investigated, its attributes, and their observations about its use in origami.

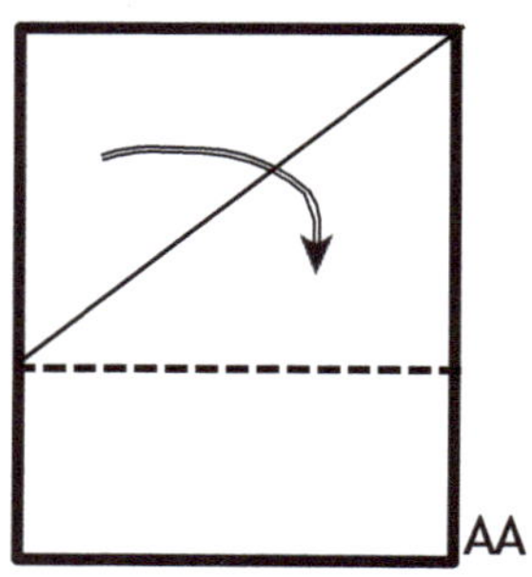

ELA Compile student journal entries into a group guide.

SQUARE ONE
STEMVESTIGATION: PAPER SQUARES

Make your own square papers for origami. Learn to fold a shape.

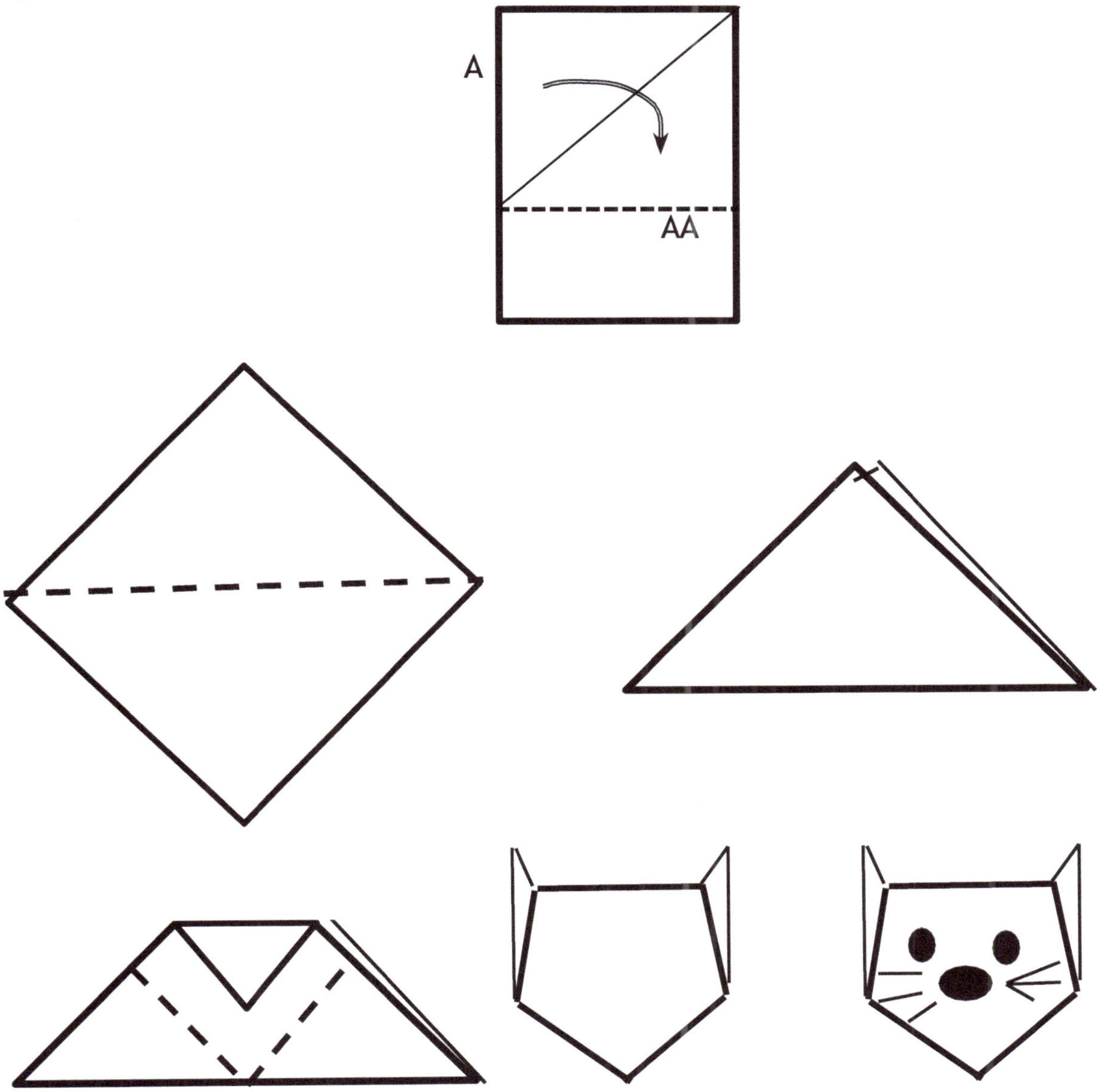

http://www.origami-instructions.com/origami-for-kids.html

Paper Forests

Explore origami in mathematics

Paper Forests

Explore origami in mathematics

Rhombi's Grandpa folded beautiful paper art. When Rhombi was little, Grandpa could make animals and flowers and boats and castles appear from his hands using nothing but squares of thin paper. Grandpa taught her to fold paper boxes to hold paper jewels. He taught Rhombi to make beautiful princesses with swords that rescued dragons from wicked wizards. Rhombi's favorite project was a mobile hung with 40 folded animals. Grandpa called it his forest in the sky.

Grandpa no longer folded paper, though. His hands were old and gnarled. The fingers couldn't hold delicate paper anymore. Now, instead of the country, Grandpa was living in a city home with lots of other grandmas and grandpas. They all played checkers and had holiday parties. Grandpa seemed happy, but sometimes he talked to Rhombi about the old days. They laughed about a funny fast squirrel that sneaked in through the window and raced around the bedroom before bursting out the front door.

Grandpa remembered pushing Rhombi on the tire swing. He talked about animals that visited from nearby woods. Grandpa especially missed the red fox that used to peer in through the bedroom windows on cold snowy nights. Sometimes they found it sleeping in his shed. Grandpa's new home was in the city. No animals peeked in city windows.

Rhombi planned to visit Grandpa for his birthday on Saturday. Sitting at her desk, Rhombi thought about a gift. She knew he missed his forest and fields, but Rhombi could not bring any live animals into Grandpa's new home. Looking at all the origami papers on her table, Rhombi decided that she could bring some animals to Grandpa after all. She picked up a stack of papers and started to fold.

Saturday was the big day. Rhombi eased her gift through the front door of Grandpa's building. Other grandmas and grandpas and aunties and uncles smiled as she made her way to the dining room. All Grandpa's friends were there with cake and balloons. Everyone oohed and ahhed as Rhombi entered. Grandpa turned to look, and his hands flew to his mouth. He smiled. He grinned.

He laughed out loud!

Hanging from Rhombi's hands was a mobile full of forest animals. His squirrel was there in its tree. There were blue birds and rabbits. But, best of all, hanging bright and sassy at the top of Grandpa's new forest in the sky was a beautiful bright red fox. Happy birthday, Grandpa!

THREE RHOMBI'S

MATERIALS

Objects with faces including circle, triangle, square, rectangle, pentagon Blindfold

SETUP

Gather shapes OR cut and laminate shapes for each group.

INSTRUCT THE PLAYERS

Students work in pairs or small groups. One student is designated "Rhombi." Rhombi will wear a blindfold and try to name the shape using only touch. Before naming the shape, Rhombi must name the number of edges and the number of vertices. Player earns a point for the complete correct answer. Rotate in a clockwise direction to the next player.

STUDENTS BREAK INTO TWO CLASS TEAMS OR SMALL TEAMS

Each team gets a pile of objects and a blindfold (baggies will keep everything neat). While students play, monitor the discussions. Players should be naming the number of edges and vertices.

UNIT 2
SQUARE ONE

Once upon a time, there was Rhombi. Rhombi loved shapes. Rhombi had been curious about circles. She had taken pictures of circles and made art with circles. She had even made a gift with circles.

She was very proud of the spinning gift she'd made for her uncle. Her designer friends had made models that helped her create the round rotating toy.

Rhombi was playing with a cardboard barn and horses. In her game, the farm's goat stomped on the barn roof. Her hands made the goat jump up and down.

Suddenly, the barn's roof cracked right down the middle. Her toy barn's roof hit the floor and broke into pieces. Rhombi was upset. She liked the barn a lot and wondered how she could fix the roof.

Rhombi knew that the roof had four edges and four corners. She put a sheet of paper in the corner to make sure they were all the right angle. She was pretty sure all the edges had been the same length. That made the roof a square. She did not know the roof's material. Rhombi had an idea.

Her friends had helped with Uncle Reel's gift. They might help her with the barn roof. Rhombi looked out from the pages of her story and asked, "Designers, will you help me create a super strong square roof for my toy barn?"

SQUARE ONE
TALK ABOUT S.T.E.A.M.

What was Rhombi's first project?
rotating gift with circles

Who liked spinning things? Hint: Rhombi gave him the toy.
Uncle Reel

What toy is Rhombi using?
toy barn

What game is Rhombi playing?
The goat is jumping on the barn roof.

What happens to the roof?
It cracks.

What does Rhombi need?
She needs a new square roof that is strong enough for her goat game.

What helps you know that a shape is a square? 4
edges, 4 corners, all edges the same length

Can you name some things that have corners like a square?
paper, ceiling and floor tiles

"Designers, will you create a square barn roof strong enough to hold 4 wood blocks?"

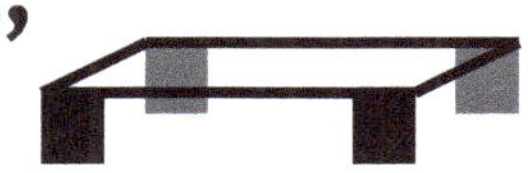

my idea	my plan

UNIT 2 RUBRIC: SQUARE ONE

Students will use their knowledge of squares to create a roof strong enough to hold 4 wood blocks.

	CONTENT	ORGANIZATION	DESIGN
1	• Is well thought out and supports the solution to the challenge or question • Reflects application of critical thinking • Has clear goal that is related to the topic • Is accurate	• Information is clearly focused in an organized and thoughtful manner • Information is constructed in a logical pattern to support the solution	• The solution is a square (taking into account age-appropriate cutting and pasting issues). • The roof holds 4 blocks.
2	• Supports the solution • Has application of critical thinking that is apparent • Has no clear goal • Has some factual errors or inconsistencies (i.e. wheels do not rotate around axle)	• Project has a focus but might stray from it at times (more concerned with form than function) • Information appears to have a pattern, but the pattern is not consistently carried out in the project	• The solution is a square (taking into account age-appropriate cutting and pasting issues). • The roof holds fewer than 4 blocks or does not support weight at all.
3	• Provides inconsistent information for solution • Has no apparent application of critical thinking • Has no clear goal • Has significant factual errors, misconceptions, or misinterpretations	• Content is unfocused and haphazard • Information does not support the solution to the challenge or question • Information has no apparent pattern	• The solution is not a square shape. • The roof does not support any weight.

RHOMBI'S POLYGON ROUND-UP
RECTANGLE

TEACHERS NOTES ABOUT THE SQUARE

- A square is a polygon with 4 sides of equal length and 4 right angle corners (90 degree corners).
- Because it has 4 sides of equal length, a square is a regular quadrilateral.
- A square is also a rectangle with equal sides and a rhombus with right angles.
- The perimeter of a square is 4 times the length of one side.
- Opposite sides of a square are parallel.
- The internal angles of a square add to 360 degrees.
- A square has 4 lines of reflectional symmetry.

UNIT 3
WE LOVE TRIANGLES

WE LOVE TRIANGLES

Essential Questions

1. How many edges and vertices are present in a regular triangle?
2. Why are triangles used in building structures?
3. Where do you find triangles at school?
4. What are you trying to accomplish in Rhombi's Challenge?

Focus on triangles, angles, regular triangles, and corners (vertices). Find apps and games to create triangles on the Internet. Locate triangles inside squares around the school. Explore primary and secondary colors. Create a triangular world as digital images are remade with triangle overlays.

Students will develop measurement skillswhiledeepeningunderstanding of 2 dimensional attributes. Connect with historical backgrounds and introduce tessellations if possible. Watch Vi Hart do some Triangle Doodling. Her videos are as entertaining as they are educational.

Help Rhombi design a fantastic new slide for a hamster using triangles as the support.

ENDURING UNDERSTANDING

Two dimensional shapes create new shapes.

The length of an object may be discussed using whole numbers.

Organizing information helps in discussion.

Number information may be organized for the purpose of answering questions.

Explaining number information makes more sense when it is organized.

Shapes may be identified by their attributes.

WE LOVE TRIANGLES
CURRICULUM CONNECTIONS

Rhombi needs to design and build a slide for the bored hamster.

MATH BACKGROUND - Triangles

A triangle is one of the basic shapes in geometry: a polygon with three corners or vertices and three sides or edges which are line segments.

TECHNOLOGY CONNECTIONS

Check out the Math Munch by Vi Hart. GOOGLE: VI HART math munch

Khan Academy's Triangle Doodling by Vi Hart. GOOGLE: khan academy triangle doodle

HISTORICAL CONNECTIONS

Triangles are present in architecture dating to earliest recorded history. Examples include:

- Egypt- Great Pyramids of Giza
- Athens- Parthenon
- Spain- Palais Ideal
- USA - Golden Gate Bridge

Unlike a rectangle, a triangle cannot be deformed without changing the length of one of its sides or breaking one of its joints. In fact, one of the simplest ways to strengthen a rectangle is to add supports that form triangles .

ART CONNECTIONS

Tilings or tessellations refers to the complete covering of a plane surface by tiles. Pompeii Tiling with equilateral triangles was an example of tessellation.

Go online to create tessellations at this site: GOOGLE: Design a tessellation online

READING CONNECTIONS

Introduce each activity and lesson with a book. Books are listed on the Read about STEAM page. Jot additional lists of books from your favorites here.

1.

2.

3.

WE LOVE TRIANGLES
CURRICULUM CONNECTIONS

COMMON CORE CONNECTIONS

ELA/Literacy

W.K.7 Participate in shared research and writing projects (e.g., explore books by a favorite author and express opinions). (K-ESS2-1)

Mathematics

MP.2 Reason abstractly and quantitatively. (K-ESS2-1)

MP.4 Model with mathematics. (K-ESS2-1)

K.CC.A Know number names and the count sequence. (K-ESS2-1)

K.MD.A.1Describe measurable attributes of objects, such as length or weight. Describe several measurable attributes of a single object. (K-ESS2-1)

K.MD.B.3 Classify objects into given categories; count the number of objects in each category and sort the categories by count. (K-ESS2-1)

NEXT GEN SCI CONNECTIONS

K-2-ETS1-1. Ask questions, make observations, and gather information about a situation people want to change to define a simple problem that can be solved through the development of a new or improved object or tool.

K-2-ETS1-2. Develop a simple sketch, drawing, or physical model to illustrate how the shape of an object helps it function as needed to solve a given problem.

K-2-ETS1-3. Analyze data from tests of two objects designed to solve the same problem to compare the strengths and weaknesses of how each performs.

K-PS2-1. Plan and conduct an investigation to compare the effects of different strengths or different directions of pushes and pulls on the motion of an object.

K-PS2-2. Analyze data to determine if a design solution works as intended to change the speed or direction of an object with a push or a pull.*

UNIT 3 S.T.E.A.M. ACTIVITIES FOR "WE LOVE TRIANGLES"

Rhombi Story Book

Rhombi Audio Download / Video (Available December 2017

UNIT 3:	We Love Triangles:	Science:	Create human pyramids.
UNIT 3:	We Love Triangles:	Technology:	Explore symmetry online.
UNIT 3:	We Love Triangles:	Engineering:	Reshape the world with triangles.
UNIT 3:	We Love Triangles:	Art & Music:	Explore primary colors with triangles.
UNIT 3:	We Love Triangles:	Mathematics:	Build strong triangle ramps.

UNIT 3: We Love Triangles: ELA Read "We Love Triangles." Students illustrate the story.

WE LOVE TRIANGLES

BUILD S.T.E.A.M. WITH GREAT BOOKS

SCIENCE

Showdown at the Food Pyramid
By Rex Barron

Triangles (Shapes in Math, Science and Nature)
By Catherine Sheldrick Ross and Bill Slavin

TECHNOLOGY

The Wing on a Flea
By Ed Emberley
GOOGLE : youtube, emberley ed

Triangles and Pyramids
By Sally Morgan

"The Polygon Song"
By Steve Saunders
GOOGLE : youtube, saunders

ENGINEERING

Triangles Around Town
By Nathan Olson

The Bridge
By Emily Cheney Neville

The Art of Construction
By Mario Salvadori

ART & MUSIC

Triangle for Adaora: An African Book of Shapes
By Ifeoma Onyefulu

Draw a Triangle, Draw Anything!
By Chris Hart

MATH

Triangles
By Bill Slavin, Catherine Sheldrick Ross

Mouse Shapes
By Ellen Stoll Walsh

ELA

The Greedy Triangle
By Marilyn Burns

Pop's Bridge
By Eve Bunting

SCIENCE

WE LOVE TRIANGLES
STEMVESTIGATIONS: TRIANGLES ARE POLYGONS

READ **Traveling Triangles**

DISCUSS Where do you see triangles?

EXPLAIN Polygons may have three, four, five, six, or more sides. All sides must be straight. Triangular polygons are more stable than other polygon shapes. The angles in a triangle cannot change once the triangle is built. The angles are fixed. Because a triangle has three sides and three angles, each angle is fixed by the side opposite to it.

MOVE Students lie on the ground to form triangles. Snap digital photos to show that height of student changes the shape of the triangle created.

INVESTIGATE

Create string and chalk triangles on the floor and pavement squares. Locate squares or rectangles on sidewalks or hall floors. Two students work together. One student holds the string from corner to corner. The second student traces the string with chalk to create two triangles.

Count triangles created by the entire class.

ELA Draw the shapes created on pavement, and explain the process to another student.

WE LOVE TRIANGLES
STEMVESTIGATIONS: HUMAN TRIANGLES

Trace these triangles.

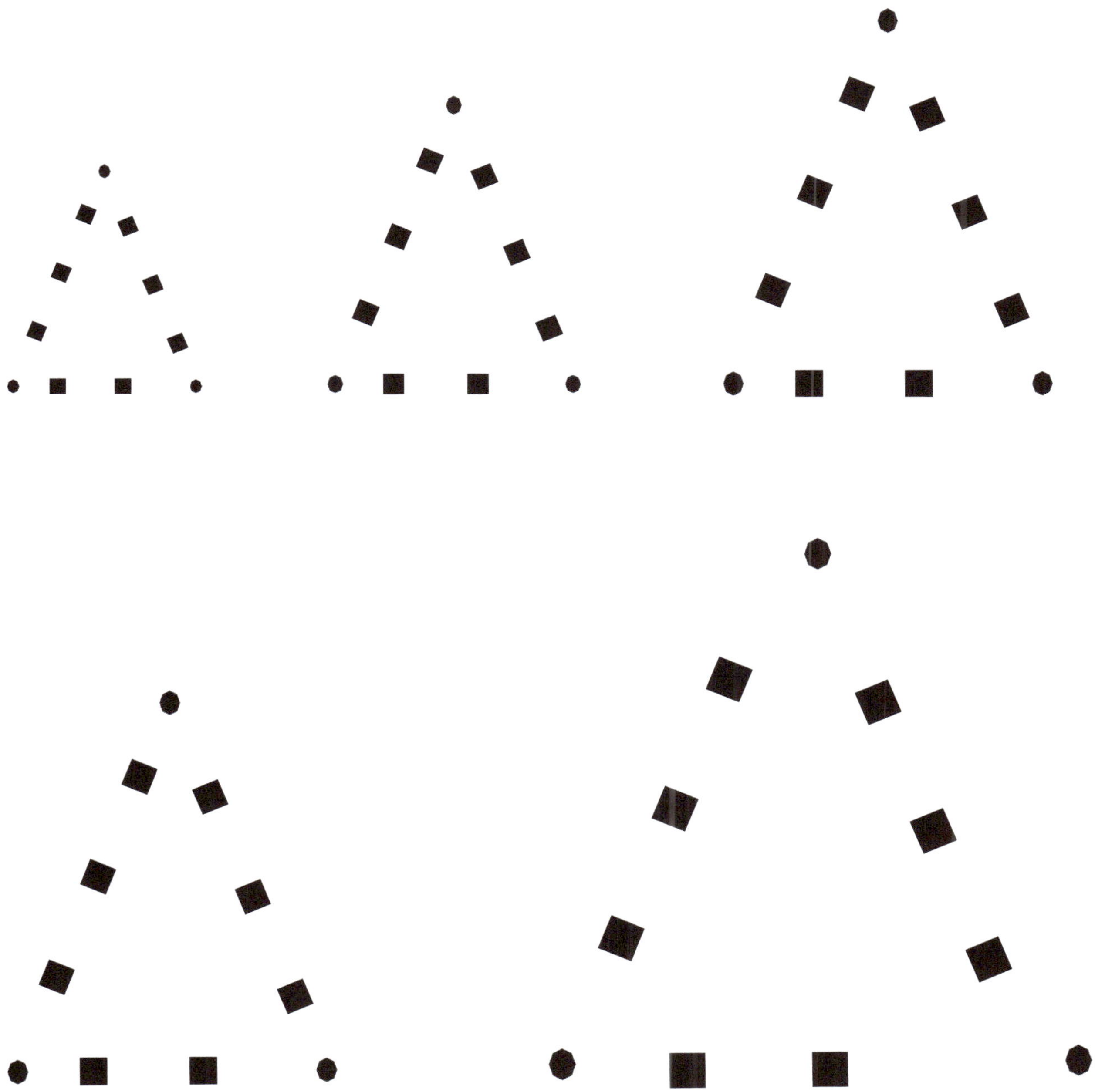

Traveling Triangle

Explore triangles in engineering

Traveling Triangle

Explore triangles in engineering

Triangle, Triangle,
where have you been?
I've soared in the sky
blown East and West by the
wind.

Triangle, Triangle,
what have you seen?
I've watched the city
from the top of a Ferris wheel.

Triangle, Triangle,
what have you done?
I've held the wood steady for
builders in the hot summer sun.

Triangle, Triangle,
where have you gone?
I've biked along city streets
in the cool quiet dawn.

Triangle, Triangle,
what should we play?
Let's climb on the jungle
gym at the end of the
day.

TECHOLOGY

WE LOVE TRIANGLES
STEMVESTIGATIONS: SYMMETRY

READ **Butterfly, Flutter By**

DISCUSS The poem is symmetrical. Each side is a mirror image. What happens in each line? Where is the mirror?

EXPLAIN Reflection symmetry happens when one half of a shape is the reflection of the other half. The line where the shape reflects is called the line of symmetry. Basically, it is where the shape's fold would happen.

MOVE Play "Rhombi Says" as you have students repeat actions on their left and right sides. They are "out" when the action is not repeated on both sides. For example, Rhombi says, "Wave your hands" and students must wave right and left hands. Rhombi says, trace a circle in the air or tap your knee or touch your nose.

INVESTIGATE

Symmetrical Poetry
The Little Boy and the Old Man, Shell Silverstein on Goodreads
http://alturl.com/p6uhx

Make Symmetrical Pictures at Kali
http://www.scienceu.com/geometry/handson/kali/

Draw the other side of each symmetrical shape.

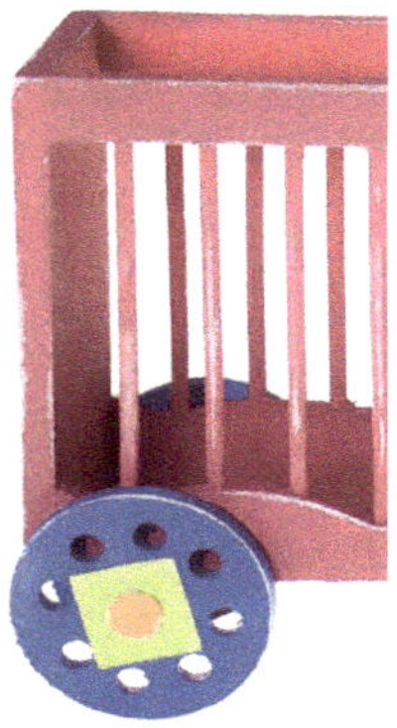

Butterfly, Flutter By

Explore symmetry

Butterfly, Flutter By

Explore symmetry

She eats candy. Candy eats she.
He loves chocolate. Chocolate loves he.

They share sunshine. Sunshine share they.
They watch butterflies.
Butterflies watch they.

Butterflies flutter. Flutter
butterflies. Butterflies are shy.
Shy are butterflies.

Wings move swiftly. Swiftly move
wings. Goodbye butterflies.
Butterflies goodbye.

ENGINEERING

WE LOVE TRIANGLES
STEMVESTIGATIONS: RESHAPE THE WORLD

READ **Tri, Tri Again**

DISCUSS The squirrels are eating food meant for the birds. What solutions are tried? What solution finally works? How would you solve the problem?

EXPLAIN Geometric shapes used in mathematics have very specific rules to keep all mathematicians on the same page. Circles are measured from the center to outer edges. Squares have four edges and four corners, and all edges must be the same length. All corners must be the same angle (like the corner of a sheet of paper).

MOVE Take a digital shape walk. Snap photos of various scenes around the school that look interesting shape-wise. You will use these images in the Investigation.

INVESTIGATE

Use the digital images students found around the school. Choose a few images to use as your foundation. Choose a few for students to use. In a presentation software program, have students place shapes over parts of the picture.

For example, place a triangle over the angled rooftop. As each group takes a turn with the image, they will add their own shape replacements. Eventually, your class will create a fully re-shaped image!

ELA Explain how the class used the software.

WE LOVE TRIANGLES
STEMVESTIGATION: RESHAPE THE WORLD

Place polygons over the picture to make a new shape.

Euclid's Books

Explore the history of geometry

Euclid's Books

Explore the history of geometry

FAQ

Euclid (YOU-klid) was one of the earliest scientists. He wanted to use reason and logic to prove the truth of things.

- Copies of Euclid's books can be found in libraries today.
- The books begin with definitions of a point, a line, and shapes.
- Euclid used geometry to prove that all right angles are equal. That means all the angles that make up corners of paper or corners of squares are shaped the same.
- Euclid taught even more interesting math. He wrote about how triangles and circles work.
- He also showed that flat shapes could be put together to make not-flat shapes. For example, six flat squares can make a cube.
- Euclid is famous because his writing was easy to understand. For two thou-sand years, Euclid's books were used as the math book around the world.
- Some mathematicians still working today started by studying geometry from Euclid's books.

ART

WE LOVE TRIANGLES
STEMVESTIGATIONS: PAINT THE TRIANGLE

READ **Paint the Triangle**

DISCUSS How many objects can be drawn using triangles? Where do you see triangles in faces? Birds? Lizards?

EXPLAIN The triangle does not exist as a regular polygon on animals. Triangles are something we've created, but we can recognize the essential shape using our imaginations.

MOVE Laminate a triangle (and other shapes). One team of students should affix the shape to an object in the classroom that contains a triangle. Other teams attempt to locate the shape first. Play progresses as teams take turns placing the triangle (and other shapes) until reaching 180 points. 180 is the total sum of all three angles inside a triangle.

INVESTIGATE

There are three primary paint colors (blue, red, yellow), the three pigments that form all art. Photocopy a regular equilateral triangle for each student. Paint one edge blue. Paint one edge red. Paint one edge yellow. Investigate the blending of primary colors. Using a paintbrush, gently pull some of each color toward the triangle's center. What colors result from blending all three? These colors are called secondary colors.

ELA Make lists of objects in each color group. Describe emotions that might correspond to the various colors.
Read Shel Silverstein's "The Giving Tree" aloud the class to act the story out in small groups.

Paint the edges of your triangle in primary colors.

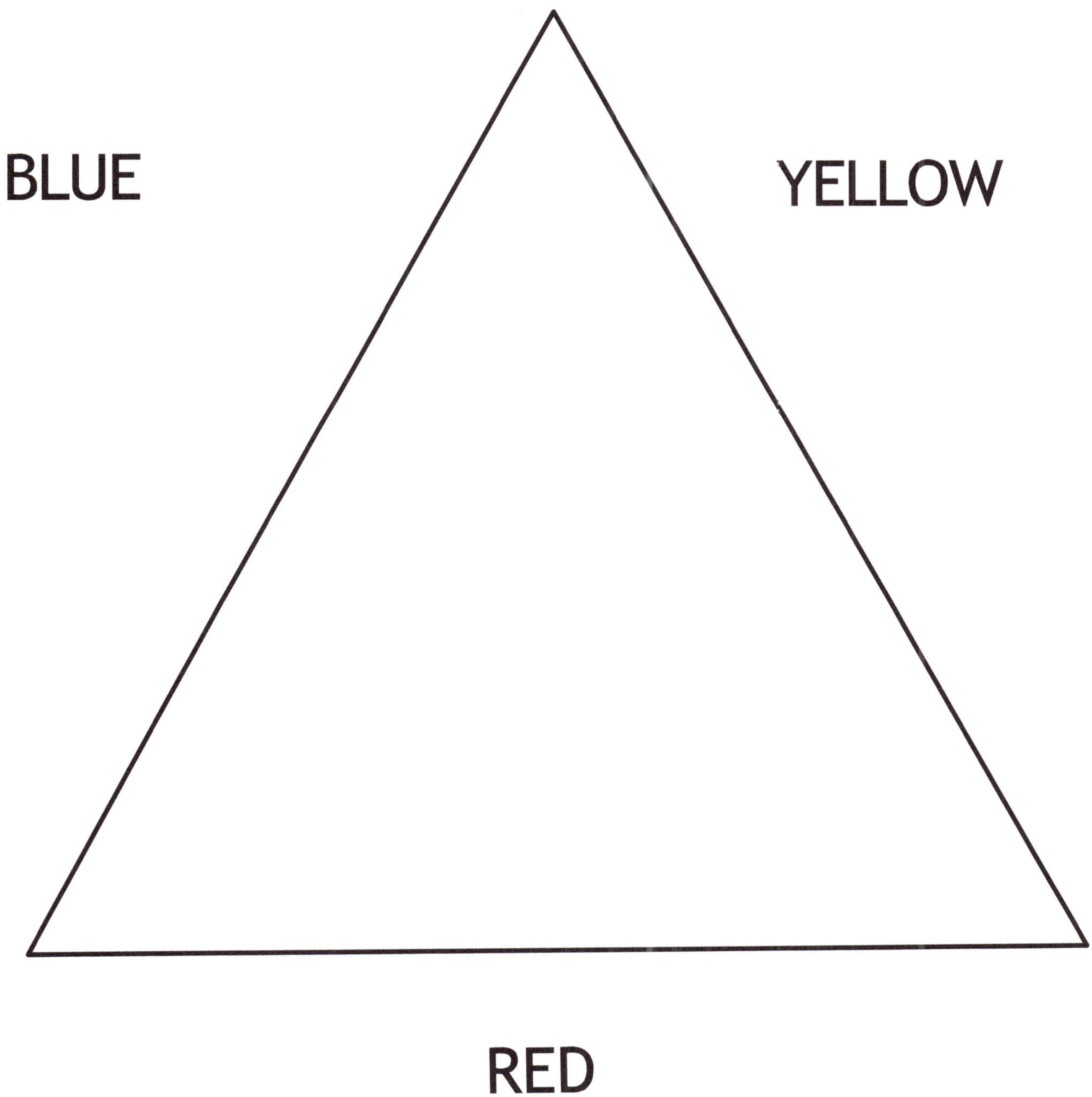

City of Angles

Explore maps in technology

City of Angles

Explore maps in technology

Every street in Three Angles was named after the shape of its corner. That made walking through town easy if you knew your shapes. Quad wanted to find Polygon Museum. He was new to the city, but he was pretty sure he could get around. Quad didn't bother taking a map.

Quad walked North along Line Street to the town square. He turned right at the square and walked East along Right Angle Road. Quad bought a cool dog collar at the Pet Palace. At the farthest end of Right Angle Road, Quad took a sharp right turn and headed South along Hypotenuse Street. He found two great bookstores along the way. Finally, Quad ended up right back where he'd started! He had not found the museum. He decided to try again on Friday.

The next morning, Quad looked for the Museum again. This time, he started on Hypotenuse Street. He headed North on Hypotenuse until the street came to a dead-end. Quad decided to turn right onto Acute Avenue.

He was sure to find the museum at the end of Acute Avenue. Quad reached the corner and took a slight right onto Obtuse Place. He walked and walked until he saw the sign for his hotel at the end of Obtuse Place just before Hypotenuse.

Good grief! The sun was setting. He'd have to find the museum on Saturday. This time, he'd save time and use a map.

On Saturday morning, Quad stood outside his hotel. The young tourist looked at his map. He touched the name of the Polygon Museum on the map. Slowly, Quad looked left. The museum was beside his hotel. Had Quad turned left out of the hotel, he would have seen the Museum right away. Next time, Quad would definitely use a map.

MATHEMATICS

WE LOVE TRIANGLES
STEMVESTIGATIONS: RAMP IT UP

READD **Euclid's Books**

DISCUSS What was Euclid? What is a mathematician? Why is Euclid still famous? He wanted to use logic to prove how triangles, lines, points, and other shapes worked. He also discussed the way flat shapes could be combined to make not-flat shapes.

EXPLAIN Triangles can make a structure stronger. You'll see triangles in bicycles, chairs, tables (look underneath), bridges, and cardboard.

MOVE Look for examples of triangles making structures strong. Students play a version of "musical chairs" using triangles in objects. Play music while students look for examples. Stop music. Anyone left without an example must sit down...

INVESTIGATE

Challenge students to make ramps from paper triangles. How will they make the ramp strong enough to support a 1:64 scale toy car (the smallest ones)?

Use extra triangles glued together to make a stronger wall.
Use popsicle sticks or straws to make the walls stronger.
What other ideas do kids have?

ELA Look at the school's ramps. Interview the administrators or custodians about why ramps are important at the school.

WE LOVE TRIANGLES
STEMVESTIGATION: RAMP IT UP

Color the car, and cut it out for your ramp.

Tri, Tri Again

Explore triangles in mathematics

Tri, Tri Again

Explore triangles in mathematics

Poly really needed to fix her bird feeder's roof. The feeder hung from a branch outside her window. All kinds of feathered friends relied on the feeder for their winter meals. Sadly, there were a bunch of furry friends that also wanted to eat at the hanging seed source. Long fluffy tails knocked the feeder to the ground over and over again. Poly was afraid that the feeder would break in one of its squirrelly falls.

Poly had tried lots of different ways to trick the squirrels. She wanted to keep them away from the feeder, but the mammals wanted all the free seeds. Squirrels had nuts of their own, but the bird feeder was such an easy food source. Poly had designed a round tipsy cover that should have made the squirrels wobble and fall to the ground. They learned how to climb around it in less than a week. Poly had turned a mesh scarf into a wrap through which squirrels couldn't get their hands. They'd just untied and dropped the rect-angular scarf to the ground. Poly had even covered the bird feeder hanger with drops of slippery oil. The greedy beasts just slid down it like a firefighter's pole. They even seemed to have fun!

Poly had thought she was out of ideas. Round plates hadn't worked. Rectangular scarves hadn't worked. Spheres of oil hadn't worked. That afternoon, on her class trip to the city mu-seum, Poly had seen a tall pointed roof. The roof covered a bell, and the guide told her class that the roof kept pigeons from landing. Bingo! If it kept pigeons away, the roof should keep squirrels away.

Poly cut out 3 triangles of wood. She covered them with waterproof material used for rain-coats and connected all three triangles at their long edges. She formed a tall pointed object and attached it to the bird feeder. Her brand new triangular roof was nearly 3 feet tall and very steep. She hung the modified feeder and sat down to wait.

Birds flew in and out to get their seed meals. Finally, two fat little squirrels came to sit on the branch above the feeder. They tilted little heads to one side and the other side looking confused. Finally, one brave little friend leaped to the roof. She slid right past the seed and landed on the ground. That was one grumpy squirrel. Her rodent friends each tried the feeder, but all failed.

Poly had learned her lesson.

If at first you don't succeed with circles and rectangles, tri, tri again.

SHAPE MIMES

MATERIALS

Pictures of shapes or shape names written on paper

SETUP

Break students into teams of 3 or 4 students. Place shape pictures or names in a bag or hat. Create a points tally chart on the board.

INSTRUCT THE PLAYERS

Teams will compete against one another for points OR adds points team-by-team to the whole class goal. Choose team names. One person from the first team chooses a shape picture / name from the bag. The team members must convey to students which shape they've picked BUT MAY NOT SPEAK AT ALL.

- Point to an example in the room.
- Make the shape with your arms.
- Form the shape with your bodies. (3 kids at the vertices of a triangle, etc.)
- Trace the shape in the air.

UNIT 3
WE LOVE TRIANGLES

Once upon a time there was Rhombi. Rhombi loved shapes.

Circles were cool. There were circles in the word "cool." Rhombi sat in the park. She looked around the playground and played her own game of "I SPY" with shapes.

"I spy with my little eye a square tile."
"I spy with my little eye 80 circles."

80 circles were linked to make a chain that held the swing. She had counted them one day at recess.

The basketball backboard was a square. She knew because each seat had 4 equal sides and 4 corners shaped like the capital letter L.

Rhombi saw triangles in the swing. The triangles held up the swingset and kept it from tipping. Rhombi decided that the triangle must be pretty strong. She watched some kids swinging really hard, but the triangles did not fall. Rhombi saw triangles in the slides and park bench, also.

After recess, Rhombi's class had center time. Rhombi went to the hamster cages to clean the cedar. She would refill the water, too. The hamsters sat in their homes. The habitats were clear plastic so kids could see the hamsters.

The poor rodents looked bored inside their 3D home. Kids used recess to recharge their brains. Rhombi wished she could give the pets recess.

Rhombi smiled and said,"Well, if I can't take you to recess. I can bring recess to you!" Rhombi looked out the window. She saw swings on the playground. Then Rhombi saw the slides.

Rhombi knew she would need some help. She looked out from the pages of her world and asked, “Designers, will you help me design and make a slide with a sturdy triangle base for our hamster? He weighs about as much as a beanbag.”

WE LOVE TRIANGLES
TALK ABOUT S.T.E.A.M.

What other circles can be found in the paragraph?
once, upon, Rhombi, loved, cool, word, also, out, how, to, for, lot, of, fun

How might Rhombi test a shape to see that it is a square?
Look for a single right angle corner. Measure all sides: they should be equal.

Where does Rhombi see triangles on the playground?
Swing, Slide, Bridge

Why doesn't the swing set tip over someone swings?
The swing uses triangles to make a stronger base.

What shape makes the bridge strong enough to hold kids?
Triangles

What does the class do right after recess?
Center time

What is Rhombi's job at today's center?
She is caring for the pet.

What pet does the class have?
hamster

What is the shape of the hamster's habitat?
cube

What project does Rhombi want to begin?
She wants to make a slide for the hamster.

Designers, will you create a hamster slide?

my idea

my plan

RHOMBI'S ADVENTURES IN 2D

UNIT 3 RUBRIC: WE LOVE TRIANGLES

TASK **Students will use their knowledge of triangles to troubleshoot and build a playground slide that holds 6 ounces of weight.**

	Content	Organization	Presentation
1	• Is well thought out and supports the solution to the challenge or question • Reflects application of critical thinking • Has clear goal that is related to the topic • Is accurate	• Information is clearly focused in an organized and thoughtful manner • Information is constructed in a logical pattern to support the solution	• Format enhances the content • The slide holds a beanbag without collapsing during repeated use. • The slide features triangles in its design.
2	• Supports the solution • Has application of critical thinking that is apparent • Has no clear goal • Has some factual errors or inconsistencies (i.e. wheels do not rotate around axle)	• Project has a focus but might stray from it at times (more concerned with form than function) • Information appears to have a pattern, but the pattern is not consistently carried out in the project	• Format does not suit the content • The slide holds a beanbag for a short time but collapses under repeated use. • The design includes triangles but does not rely on it for strength.
3	• Provides inconsistent information for solution • Has no apparent application of critical thinking • Has no clear goal • Has significant factual errors, misconceptions, or misinterpretations	• Content is unfocused and haphazard • Information does not support the solution to the challenge or question • Information has no apparent pattern	• Presentation appears sloppy and/or unfinished. • The slide collapses under any weight at all. • There are no triangles included.

ABOUT THE TRIANGLE

- Triangles are polygons with the least possible number of sides (three).
- The three internal angles of a triangle always add to 180 degrees.
- An equilateral triangle has three sides of equal length and three equal angles.
- The longest side of a right angle triangle is called the hypotenuse, it is always found opposite the right angle.
- Trigonometry is the study of the relationship between the angles of triangles and their sides.
- Triangle shapes are often used in construction because of their great strength.

UNIT 4
TAKE FIVE

Essential Questions

1. How many edges and vertices are present in a regular pentagon?
2. What important monument is build in the shape of a pentagon?
3. How would you draw a pentagon?
4. What are you trying to accomplish in Rhombi's Challenge?

ENDURING UNDERSTANDING

Two dimensional shapes create new shapes.

The length of an object may be discussed using whole numbers.

Organizing information helps in discussion.

Number information may be organized for the purpose of answering questions.

Shapes may be identified by their attributes.

Symmetry means a shape can be folded so edges will match

Asking questions and taking part in discussions increases understanding of a subject.

TAKE FIVE

How many different objects, flora and fauna exist in a pentagonal shaped piece of land? What inventions would help the class? Design and improve signage material. Learn to read infographics. Investigate and create comic strips with 5 panels. Discover the true size of America's famous pentagon.

Students will count to 100 and bundle, compose and decompose numbers. Halves, wholes, and quarters are introduced in folding paper to make pentagons. Discover symmetry and deepen measurement and geometric understanding.

Help Rhombi update the hiking trail signage with sturdy, brightly colored deer crossing signs.

TAKE FIVE
CURRICULUM CONNECTIONS

Rhombi needs to build deer crossing signs.

MATH BACKGROUND - Pentagons

A pentagon is a 5 sided polygon with interior angles that add to 540 degrees. Regular pentagons have edges/sides of equal length and interior angles / vertices of 108 degrees. The edible plant okra is shaped like a pentagon.

TECHNOLOGY CONNECTIONS

Julett Morris Elementary Explains Shapes
http://www.schooltube.com/organization/122887/
Students use shapes to build objects.
http://bit.ly/g8OY97
Shapes Song by Peter Weatherall
http://www.youtube.com/user/peterweatherall/videos

HISTORICAL CONNECTIONS

The Pentagon is home to Department of Defense and has three times the floor space of the Empire State Building in New York. According to the website, "Despite 17.5 miles of corridors it takes only seven minutes to walk between any two points in the building." Pentagon Tours explain that there "are 19 escalators, 131 stairways, 284 restrooms, 691 water fountains and enough telephone wire in the building to wrap around the world four and one-half times!"

ART CONNECTIONS

Draw a pentagon with pencil and ruler. Use the straight edge to draw a line from corner to corner diagonally. Repeat in the opposite corners. Draw a dot in the center.

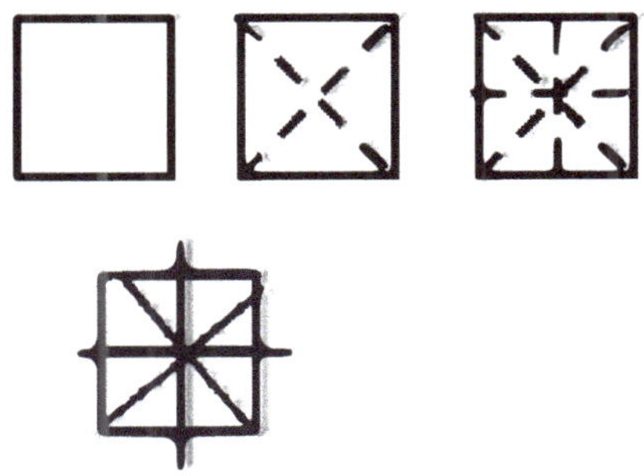

READING CONNECTIONS

Introduce each activity and lesson with a book. Books are listed on the Read about STEAM page. Jot additional lists of books from your favorites here.

1.

2.

3.

TAKE FIVE
CURRICULUM CONNECTIONS

COMMON CORE CONNECTIONS

ELA/Literacy

W.K.7 Participate in shared research and writing projects (e.g., explore books by a favorite author and express opinions). (K-ESS2-1) add Venn

Mathematics

MP.2 Reason abstractly and quantitatively. (K-ESS2-1)
MP.4 Model with mathematics. (K-ESS2-1)
K.CC.A Know number names and the count sequence. (K-ESS2-1)
K.MD.A.1Describe measurable attributes of objects, such as length or weight. Describe several measurable attributes of a single object. (K-ESS2-1)
K.MD.B.3 Classify objects into given categories; count the number of objects in each category and sort the categories by count. (K-ESS2-1)

NEXT GEN SCI CONNECTIONS

K-2-ETS1-1. Ask questions, make observations, and gather information about a situation people want to change to define a simple problem that can be solved through the development of a new or improved object or tool.
K-2-ETS1-2. Develop a simple sketch, drawing, or physical model to illustrate how the shape of an object helps it function as needed to solve a given problem.
K-2-ETS1-3. Analyze data from tests of two objects designed to solve the same problem to compare the strengths and weaknesses of how each performs.

UNIT 4 S.T.E.A.M. ACTIVITIES FOR "TAKE FIVE"

Rhombi Audio Download and Videos Available April 2017

UNIT 4: Take Five: Science: Take part in a pentagon treasure Hunt
UNIT 4: Take Five: Technology: Can students design a new tool?
UNIT 4: Take Five: Engineering: Design and improve signage material.
UNIT 4: Take Five: Art & Music: Investigate and create comic strips with 5 panels.
UNIT 4: Take Five: Mathematics: Discover the true size of America's famous pentagon.

UNIT 4: Take Five: ELA Activity: Read Rhombi's "TAKE FIVE." Student's illustrate the story.

TAKE FIVE

BUILD S.T.E.A.M. WITH GREAT BOOKS

SCIENCE

Grand Canyon
By Cari Meister

One Small Square (caves)
by Donald Silver

Shapes in Sports
By Rebecca Rissman

TECHNOLOGY

Goodnight, Goodnight Construction Site
Sherri Duskey Rinker

Tool, Time, Twist: A Brief History of Tools Through Time
By David Shapiro

http://www.trafficsign.us

ENGINEERING

Monkey with a Tool Belt
By Chris Monroe

The Kids' Building Workshop
By Craig Robertson

Shapes in Transportation
By Jennifer Rozines/ Roy Roy

ART & MUSIC

13 Art Inventions Children Should Know
By Florian Heine

Shapes (Museum of Modern Art)
by Philip Yenawine

Math-terpieces By
Greg Tang

MATH

Pentagon
By Sheila Rivera

How Long or How Wide?: A Measuring Guide
By Brian P. Cleary

ELA

To Root to Tootto Parachute: What Is a Verb
By Brian P. Cleary

How Margaret E. Knight Became an Inventor
By Marvelous Mattie
Emily Arnold McCully

SCIENCE

TAKE FIVE
STEMVESTIGATIONS: PENTAGON TREASURE HUNT

READ **We're Hunting for Pentagon**

DISCUSS What other rhyming game uses this pattern? What pentagons do we find in the game? Can you think of any other pentagons you may have seen?

EXPLAIN Pentagons are rare in nature and in human structures. A house, a crossing sign, a morning glory, and the pattern on a soccer ball are mentioned.

MOVE Clap and act out "We're Hunting for a Pentagon."

INVESTIGATE

Stake out one large class-sized pentagon in the grass. Give students magnifying glasses and journals. Students record what they find in the tiny habitat.

Insects
Flowers
Weeds
Rocks, Dirt, or Soil
Trees
Trash or cast-off wrappers, etc.

ELA Share findings with the class.

TAKE FIVE
STEMVESTIGATION: STAKE OUT THE PENTAGON

What lives on a pentagon of Earth? Draw what you see.

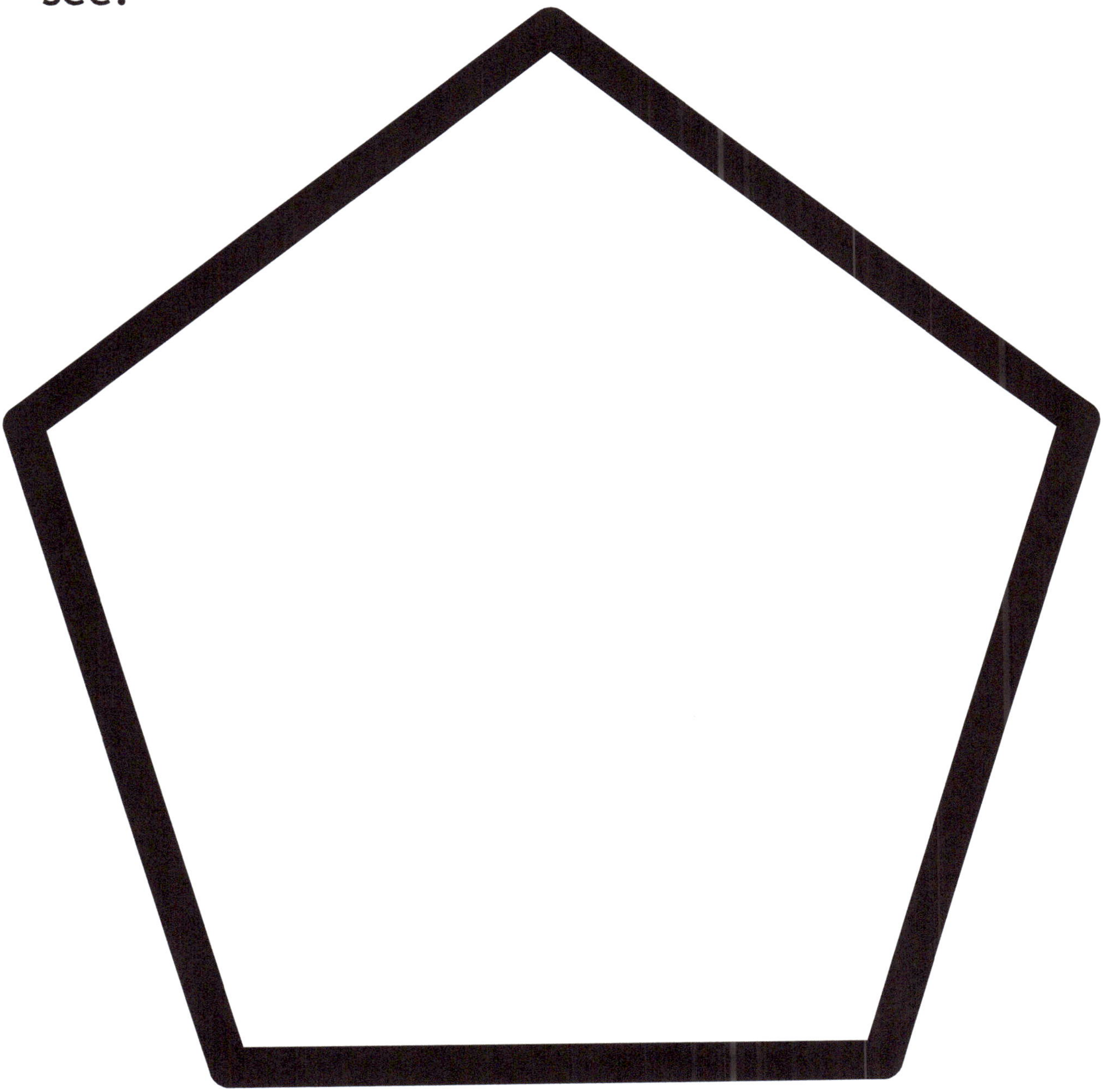

We're Hunting for a Pentagon

Explore pentagons in science

We're Hunting for a Pentagon

Explore pentagons in science

We're going on a 5 hunt. (We're going on a 5 hunt.)
We're looking for a pentagon. (We're looking for a pentagon.)
We're gonna catch a big one. (We're gonna catch a big one.)
I'm not afraid. (I'm not afraid.)
Are you? (No way!)

Here comes the crossing sign.
Better wait for the crossing guard.
Walk briskly and pump your arms.

REFRAIN

We're passing a tall house. It sure is wide and has 5 edges. Let's walk around it.

REFRAIN

We're walking through the soccer field. Uh-oh. Here comes the soccer ball. We better duck under it...

REFRAIN

We're running through the garden now. It sure is grassy. I see the morning glories and okra. Jump over the rows. Oooohhh. We're all muddy. Shake it off....

REFRAIN

I see a big building. It has 5 sides. It's The Pentagon.
Oh no, what time is it?
Recess starts in 5 minutes. We're going to miss it.
Better get home fast!
Oh no, mud! Shake it off.
Jump the okra and morning glories.
Yikes, duck the soccer ball.
Walk quickly around the tall house.

STOP! The crossing guard says to wait. Okay, we can cross now.
Walk briskly and pump those arms. Hurry! Hurry!
Ahhhhh, we made it!

TECHOLOGY

TAKE FIVE
STEMVESTIGATIONS: TOOLS

READ Tool Time

DISCUSS Tools are non-consumable physical objects used to achieve a goal. Tool use has been documented since the beginning of human history, and animals also employ tools. Astronomers use telescopes to see faraway places in greater detail.

EXPLAIN Humans have created tools to simplify work since the stone age. Explore the history of tools from the stone hand ax to the modern toolbox. 20 tools and 2.5 million years of tool making offer a lot to think about.

MOVE Visit the web to locate great songs about construction and tools. Several fun songs that can be acted out while sung.
GOOGLE: PreSchool FUN Songs construction

INVESTIGATE

What activities could be enhanced by the use of an appropriate tool? Students should brainstorm a list of things they do each day that could be easier. What tasks are time-consuming, laborious, physically difficult, or just plain annoying?

Students should break into teams of four to design and develop a tool to achieve a goal. Identify the need. Explain the solution. Compare the idea to other existing tools. List materials to be used in the tool's creation. Gather supplies to create the tools and demonstrate their uses.

ELA Create "advertisements" to sell tool ideas to investors.

TAKE FIVE
STEMVESTIGATION: TOOLS

Tools make life easier. Draw the use for each tool.

Tool Time

Explore tools in technology

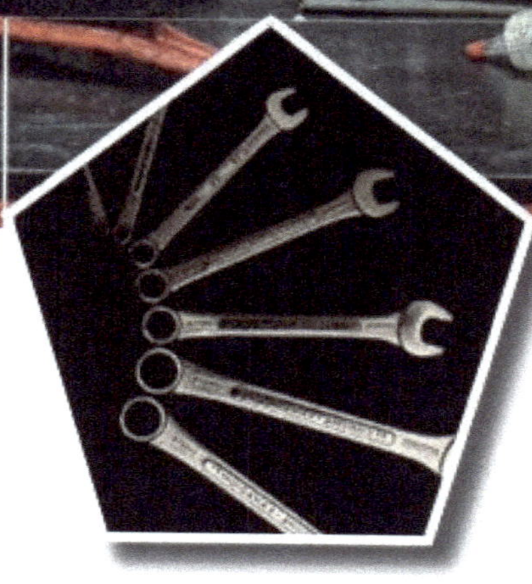

Tool Time

Explore tools in technology

Hilda and Poly had a job to do. Grandma wanted the garage organized. She liked everything in its place. The two friends had taken out the trash. They had bagged cardboard, newspaper, glass and cans for recycling. They had even swept and dusted.

The tool bench was still a big mess. Tools were scattered on the table. Tools were dumped in boxes under the bench. Tools were hung in all the wrong pegs on the wall.

Hilda and Poly started by sorting the tools used to move items. They hung hammers and mallets on the peg board over jars of nails in lots of sizes. Star-shaped philips head and flathead screwdrivers used to tighten or loosen things went into bins by size. #1, #2, and #3 were matched to their same sized screws.

Rhombi put grabbing, twisting, and pulling tools on the pegboard. Pliers were hung from smallest to largest. Wrenches were hung over jars of nuts and bolts.

Rhombi and Quad did not touch the cutting tools. Saws, hacksaws, pincers and wire cutters were left for Grandma to handle later.

Carpenter's levels, rulers, angles, squares, and other measuring tools were hung neatly to the right of the pegboard wall behind the bench. A caliper used to measure very tiny items was placed in its box. Some tools were delicate.

When Hilda and Poly finished the job, they proudly called Grandma to see her shiny garage. The excited old woman clapped her hands and hugged both girls. When Grandma offered them money for doing the job, Hilda refused to take the money.

"No, Grandma," said Hilda. "I'm happy to help!"

ENGINEERING

TAKE FIVE
STEMVESTIGATIONS: MATERIALS

READ Creative Design

DISCUSS "Pentagons" introduces shapes in a repetitive format that makes vocabulary memorable. Pentagons are most widely recognized by students in the national military building and in school crossing street signage.

EXPLAIN The pentagon is a shape with 5 sides, 5 edges, 5 corners, and 5 vertices. A regular pentagon's edges are all the same length, and its vertices are all the same angle.

MOVE Take a walk out to the "crosswalk" sign.

INVESTIGATE

Crosswalk signs are rigid pieces of metal or plastic that can stand up to strong winds, rain, and changing temperatures. How strong are other materials used in the classroom? Choose 3-4 different materials (fabrics, papers, cardboards, plastic cups, metal pans, etc.) Cut all materials to the same size pentagons.

Each team of students will make a crosswalk sign out of materials they choose. The team must explain their choices to the class.

ELA Design other school safety messages for the PENTAGONAL crosswalk sign!

Draw your pentagon crosswalk sign.

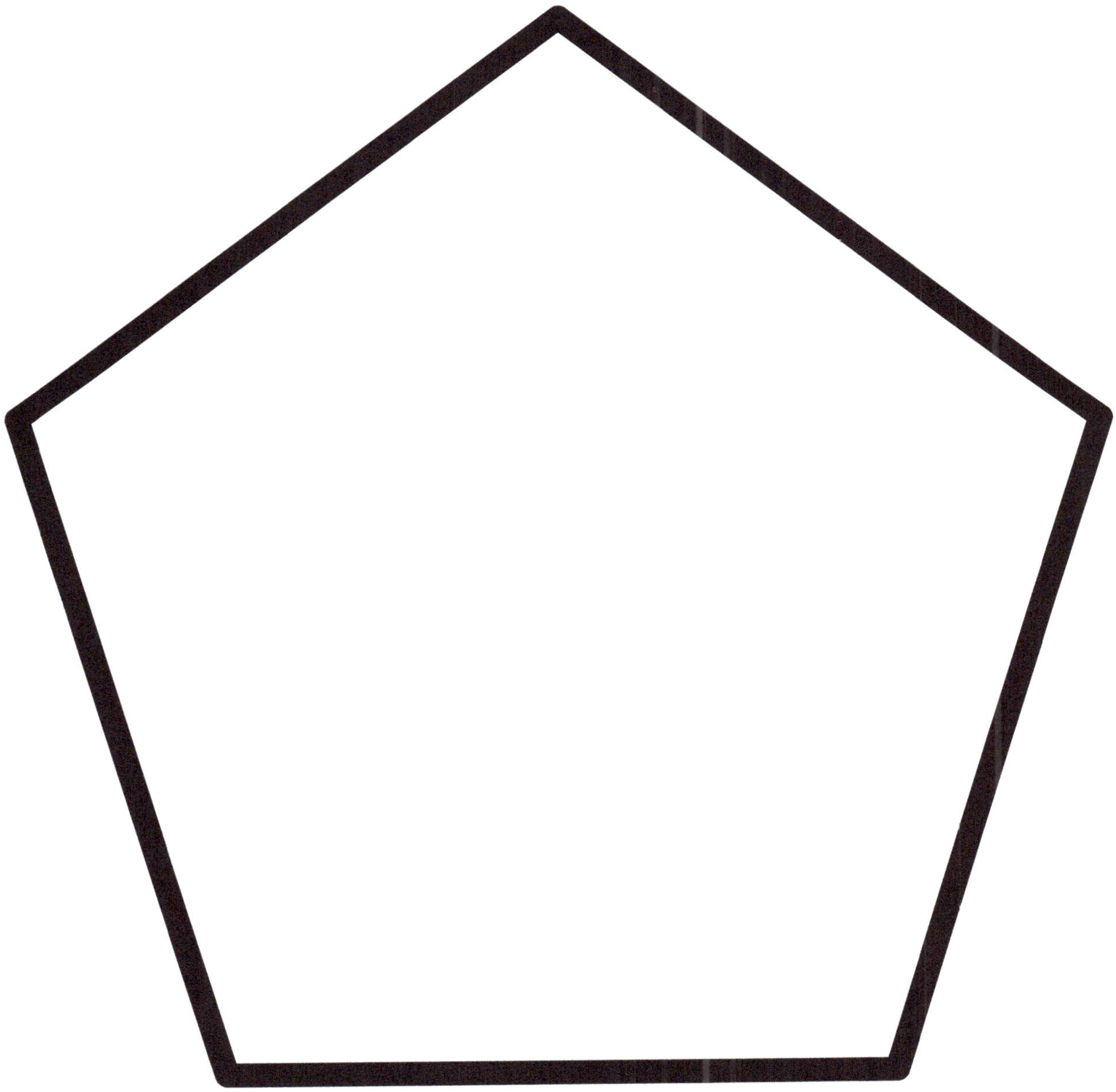

Creative Design

Explore pentagons in engineering

How Many Pentagons Can You Count In This Set Of Pictures?

Creative Design

Explore pentagons in engineering

Pentagons have 5 edges, 5 edges, 5 edges.
Pentagons have 5 edges, regular or not.

Regular pentagons have 5 edges all the same, all the same, all the same. Regular pentagons have 5 edges all the same length.

Irregular pentagons have 5 edges not the same, not the same, not the same. Irregular pentagons have 5 edges NOT the same size.

ALL pentagons have 5 angles, 5 angles, 5 angles.
All pentagons have 5 angles
regular or not.

The angles of all pentagons add up, add up, add up. The angles of all pentagons add up to 540 degrees.

5 Edges
5 Angles
500 + 40 Degrees
PENTAGONS are fun!

TAKE FIVE
STEMVESTIGATIONS: Explore Hexagons

READ Geodesic Domes

DISCUSS A humanitarian is someone who designs and creates and innovates to make life better for mankind. One of Fuller's inventions was the geodesic dome. Domes make cheap, strong houses. People are so used to the box house that his invention never became popular.

EXPLAIN Inventions take many forms. The Utah State Intellectual Properties Services defines an invention as "the discovery or creation of a new material (either a new manufactured product or a new composition or matter), a new process, a new use for an existing material, or any improvements of any of these."

MOVE Pull several comic strips from the newspaper. Students act out the panels while other try to identify the matching strip.

INVESTIGATE

Katzenjammer Kids was first printed in the New York Journal in 1897. What are students' favorite cartoons and comic strips? Cut out cartoons from the newspaper to share. Students read the cartoons (or have them read) to get ideas. What do all cartoons seem to have in common? What aspects of each cartoon are unique? Create cartoons to support a science or math lesson. Characters might -for example- have to solve a math puzzle or mystery. They might have to explain a scientific concept being learned in school.

ELA Read the story aloud in groups.

TAKE FIVE
STEMVESTIGATION: STORY IN 5 BOXES

Create a comic strip story in five parts.

TITLE

Geodesic Domes

Explore hexagons

Geodesic Domes

Explore hexagons

Young man, you amaze me. *–Albert Einstein, physicist (1938)*

Buckminster Fuller said, "Whenever I draw a circle, I immediately want to step out of it." Buckminster Fuller was an architect and an engineer born in 1895. He was also a designer and an inventor. Fuller wrote 28 books on subjects in math and science. He had 47 special degrees from different universities.

Fuller was someone that cared about making the world better. He was also a humanitarian, someone that wants to improve life for other people. Each year the Buckminster Fuller Institute offers a $100,000 prize to " scientists, students, designers, architects, activists, entrepreneurs, artists and planners from all over the world using innovative solutions to solve some of humanity's most pressing problems." From a floating doctor's office to a new plan for food production, Fuller's prize helps people with great ideas afford to put their plans in action.

This inventor was interested in using the planet's resources more efficiently. He invented a geodesic dome made of polygon shapes. The dome uses fewer materials and can be made bigger or smaller to meet the needs of many different types of uses. Gardens, homes, and museums are shaped like geodesic domes.

When Buckminster Fuller first imagined the special dome, he thought of it as humans working like a crew to survive on our planet, "spaceship earth." Epcot Center in Florida created a domed building inspired by Fuller's idea.

Domes like these are made of triangles. As more triangles are added, the shape gets stronger. No other 3-dimensional shape can be so big without extra supports inside the building. The larger it is, the stronger the dome becomes. These domes are so simple that an entire house can be built in a single day. They are so strong that geodesic domes can even survive hurricane force winds!

Can you see the triangles in the playground dome? Can you see diamonds in the playground dome? Can you see hexagons in the playground dome?

MATHEMATICS

TAKE FIVE
STEMVESTIGATIONS: THE PENTAGON

READ The Pentagon

DISCUSS The Pentagon is one of America's best known buildings. Its design was chosen to fit an unusually shaped piece of land. When the site changed, architects and engineers kept the pentagonal shape of the building for its new location.

EXPLAIN According to GreatBuildings.com, each outer edge has a length of 921feet. Other numbers include:

- Floor area of 6.5 million square feet or 34 acres
- 3.7 million square feet are used for offices
- 80 million dollar construction cost in 1943 dollars
- 17.5 miles of corridors

MOVE Each edge of the Pentagon measures 921 feet. Walk 921feet. Walk it 5 times to simulate walking the perimeter of this huge building.

INVESTIGATE

As a class, create a scale model of the Pentagon building with blocks and other classroom construction materials. If each outer wall measures 921 feet, adjust your classroom or playground structure to measure a scaled down 9.21 feet on each edge. How will the class make sure the structure is actually a regular pentagon with equal angles? (Hint: They can print and cut out a paper sized pentagon. Lay the paper at your large pentagon's center. Assemble large walls parallel to the small pentagon's edges.) What materials will you use?

ELA Rethink your comic strip. Add facts to the original comic.

STEMVESTIGATION: 5-SIDED SHAPES

Walk The Pentagon. Test its angles with a pattern.

Graphic Information

Explore infographics in mathematics

Infographic is short for information graphic. An infographic uses pictures and words to make lots of information and numbers easy to grasp. Look at this infographic showing information about The Pentagon. You could read the paragraphs that follow, or you can use an infographic to show the same information.

The Pentagon was designed in just 3 days and built far faster than most government build-ings. It was also built to hold a lot of cars. Parking was planned for 10,000 cars. 16 months passed from the first day of construction until employees parked all those cars in the lots. The construction trucks, earth movers, and builders showed up for work on September 11, 1941. The Pentagon was completed in January of 1943. The building opened almost 8 decades ago.

The building serves a lot of needs. Very special visitors are invited to parties in the formal dining room. Most armed services workers eat in one of the 2 cafeterias. 5 different armed services use the building. 10,000 or more visitors use any of the 6 snack bars.

Finding a bathroom in The Pentagon is easy. When the building was planned in 1941, Virginia state law said that black people and white people had to use different. The pentagon was built with 284 bathrooms. By order of the President, none of those bathrooms were ever really set aside for any special group of people. Employees and visitors to The Pentagon all used any of the public bathrooms. For a time, The Pentagon was the only non-segregated building in the state of Virginia.

Getting from place to place in the huge building takes time. Each outer edge measures 921 feet. There are 5 stories full of offices and a 5 acre central area called "ground zero." The total square footage of the The Pentagon is more than 5 million square feet. Your school probably measures less than 50,000 square feet. Walking from one corner to another cor-ner of The Pentagon takes nearly 7 minutes. How many seconds pass if you walk from one corner of your building to another corner of your building?

Which format seems easiest to use? Which format gives information quickest? Which format offers more of a story? Do you like the paragraphs or graphic style best?

The Pentagon, USA

HOLDS A LOT OF CARS

 = 10,000

Aerial Image by David B. Gleason from Chicago, IL (The Pentagon) via Wikimedia Commons

COUNT on the PENTAGON

1 formal dining room
2 cafeterias
3 days to design the building
4 thousand 2 hundred clocks
5 edgesandverticesinapentagon
6 snack bars
7 minute walk corner to corner
8 decades since building opened
9 hundred 21 feet per outer edge
10 thousand visitors a year

5 Stories Tall
+2 below ground

5 Armed Services
housed on site

5 Acre Central Plaza
"Ground Zero"

Executive Order 8802 let The Pentagon open all its 284 bathrooms to all employees, black or white. For a time, The Pentagon was the only non-segregated building in Virginia.

NUMBER OF MONTHS FROM GROUND BREAKING TO OPENING CEREMONY

	JAN	FEB	MAR	APR	MAY	JUNE	JULY	AUG	SEPT	NOV	DEC
1941											
1942											
1943											

Reference Material
January 15, 2013 9 Things You May Not Know About the Pentagon By Barbara Maranzani via History.com
Vogel, Steve. *The Pentagon: A History.* New York: Random House, 2007. ISBN 1-400-06303-5 (pp. 199-200, 216-217)
History of the Pentagon — U.S. DOD info, archived at Archive.org
Pentagon Web Site — By the U.S. Department of Defense.
http://en.wikipedia.org/wiki/The_Pentagon and Wikimedia Commons

DOT to DOT

MATERIALS

You will need an overhead projector or smart board setup. Copy the dot paper on a transparency. You will need appropriate marking tools (transparency markers or smart mouse setup).

SETUP

Project dot paper on a screen or whiteboard.

INSTRUCT THE PLAYERS

Taking turns, the first player connects two dots with either a horizontal or vertical line. The other players take their turns doing the same. Their lines may touch other player's lines, but they don't have to. When a player draws the fourth line to create a box, the team claims the box and writes the first letter of its team name inside the box.

STUDENTS BREAK INTO TWO CLASS TEAMS OR SMALL TEAMS

Play. Once all of the possible horizontal and vertical lines in the grid have been turned into boxes, each team counts the number of boxes it owns. The team with the most boxes wins.

UNIT 4
TAKE FIVE CHALLENGE

Rhombi, Quad and Poly put on hiking boots. They fill bottles of water. All three friends wear long pants and long sleeved shirts. Quad hands each friend some sunscreen, and they all put on a hat.

The friends will look for deer, snakes, raccoons, birds, and turtles. The kids open a map of Pentagon Park with 5 edges and 5 vertices. It's a pentagon.

"Hey, the Nature Center trails are laid out in a pentagon," Poly shouts. "That's like the shape on my soccer ball."

Quad laughs but shushes Poly. "We will never see any animals if you shout."

The kids take a trail leading to the deer meadow. With any luck -and quiet voices- they will see some animals in the woods. They hike for half an hour.

“Look at the deer,” whispers Poly.

Peeking from the edge of the woods are two deer. Many deer have used this place to cross the path. They are perfectly still and silent.

Suddenly, there is a commotion. Mountain bikers speed down the trail toward the deer. Poly shouts at the deer to move faster. She shouts at the bikers to stop!

The bikers have permits and are being safe, but an accident could happen any second! The mountain bikes skid to a stop. Gravel and dirt fly into the air. The last deer bounds away into the trees.

"There should be a sign to let us know this is a place where deer cross the path. If this is a place the deer walk a lot," says one biker. She is breathing hard.

Rhombi decides to ask for help on this project. She looks out from the pages of her world and asks, "Designers, will you create deer crossing signs like the school crossing signs outside your school?"

TALK ABOUT S.T.E.A.M.

What is another way to say
spin? circle

What innovative rooftop squares did you
create? answers vary

What shape made the hamster slide
strong? triangle

What steps do the kids take to be safe in the woods?
They wear long sleeves, long pants, hats, sunscreen and water bottles. They take maps.

What animals did the kids
see? deer

Why did the kids decide to post a "deer crossing" sign?
deer and some mountain bikers nearly ran into each other.

What is the shape of the 5-sided
sign? pentagon

Name at least 3 objects that have regular pentagons?
crosswalk sign, Pentagon building in Washington, D.C., USA, soccer ball shapes

What are the rules for a regular
pentagon? 5 equal edges, 5 equal
vertices / corners

"Designers, will you create a deer crossing sign to help the deer and mountain bikers stay safe on a trail?"

my idea	my plan

UNIT 4 RUBRIC: TAKE FIVE

TASK **Students will use their knowledge of pentagons, materials, and weather to design, build, and troubleshoot a deer crossing sign.**

	Content	Organization	Design
1	• Is well thought out and supports the solution to the challenge or question • Reflects application of critical thinking • Has clear goal that is related to the topic • Is accurate	• Information is clearly focused in an organized and thoughtful manner • Information is constructed in a logical pattern to support the solution	• Sign clearly indicates a deer crossing. • Sign meets the shape criteria of a pentagon. • Sign is brightly colored...
2	• Supports the solution • Has application of critical thinking that is apparent • Has no clear goal • Has some factual errors or inconsistencies (i.e. wheels do not rotate around axle)	• Project has a focus but might stray from it at times (more concerned with form than function) • Information appears to have a pattern, but the pattern is not consistently carried out in the project	• Sign vaguely indicates a deer crossing • Sign is a pentagon • Sign is dull and not bright enough to gain attention
3	• Provides inconsistent information for solution • Has no apparent application of critical thinking • Has no clear goal • Has significant factual errors, misconceptions, or misinterpretations	• Content is unfocused and haphazard • Information does not support the solution to the challenge or question • Information has no apparent pattern	• Project is sloppy and unfinished given child's ability level • Project does not indicate its purpose • Project does not meet minimum criteria

RHOMBI'S POLYGON ROUND-UP PENTAGON

TEACHER NOTES ABOUT THE PENTAGON

- A pentagon is a 5 sided polygon with interior angles that add to 540 degrees.
- Regular pentagons have sides of equal length and interior angles of 108 degrees.
- The US Department of Defense headquarters is named 'the Pentagon'.
- The edible plant okra is shaped like a pentagon.

EXTENSION
BEST IN SHOW

BEST IN SHOW

Use polygons in problem solving.

Once upon a time, there was Rhombi. She had very good friends that helped her design and create. Her friends had made spinning toys. They had designed innovative ways to fix the square roof on her toy barn. Friends had built a slide for the hamster from triangles. They had even designed signs to keep deer and mountain bikers safe in the woods.

Rhombi and her friends stood outside the local animal shelter. The shelter took care of pets whose owners could no longer care for them. Dogs, cats, rabbits, guinea pigs, birds, hamsters, snakes, and even a few lizards were all waiting for new families.

Rhombi, Quad, Pi, Radius, Ginny and a few other kids were planning to help the shelter animals. They hoped to make life better for the animals until every pet found its "forever home."

“We need to decide on a project,”said Rhombi.

Quad thought raising money with a bake sale was a great idea. Pi and Radius remembered adopting their dog from the shelter. The dog had been clean but scruffy. They suggested holding a pet wash. Everyone grinned at the idea of giving some angry cat a bath.

“Maybe a pet wash is not the best idea,”giggled Pi.

Pi's phone beeped. She stepped away to answer the phone. When she ended the call, Pi was jumping up and down with a huge grin on her face. Pi was waving her hands around and talking quickly. She shouted, "My aunt just called. She just bought a super cool new rabbit hutch, and it gave me an idea. What about making new kennels and cages for the animals?"

Everyone agreed that this was a great idea. They would need lots of help planning and building. Rhombi and her friends looked beyond the pages of their world to ask, "Designers, will you work in teams to help us plan innovative new habitats for the animals in the shelter?"

Designers, will you create a new kind of pet house?

PYRAMID NET

INSTRUCTIONS: Print this net for use in a center or during STEMVESTIGATIONS.

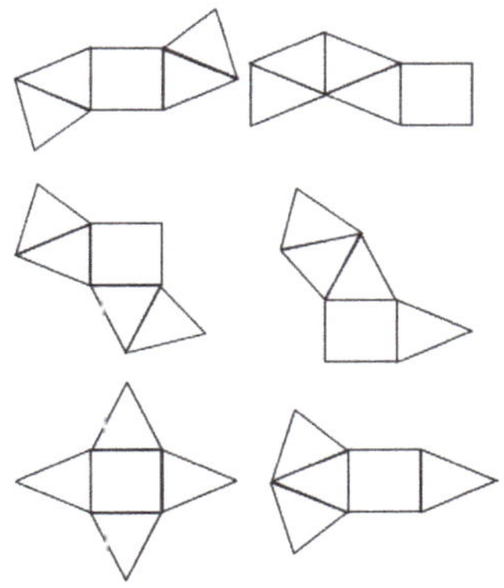

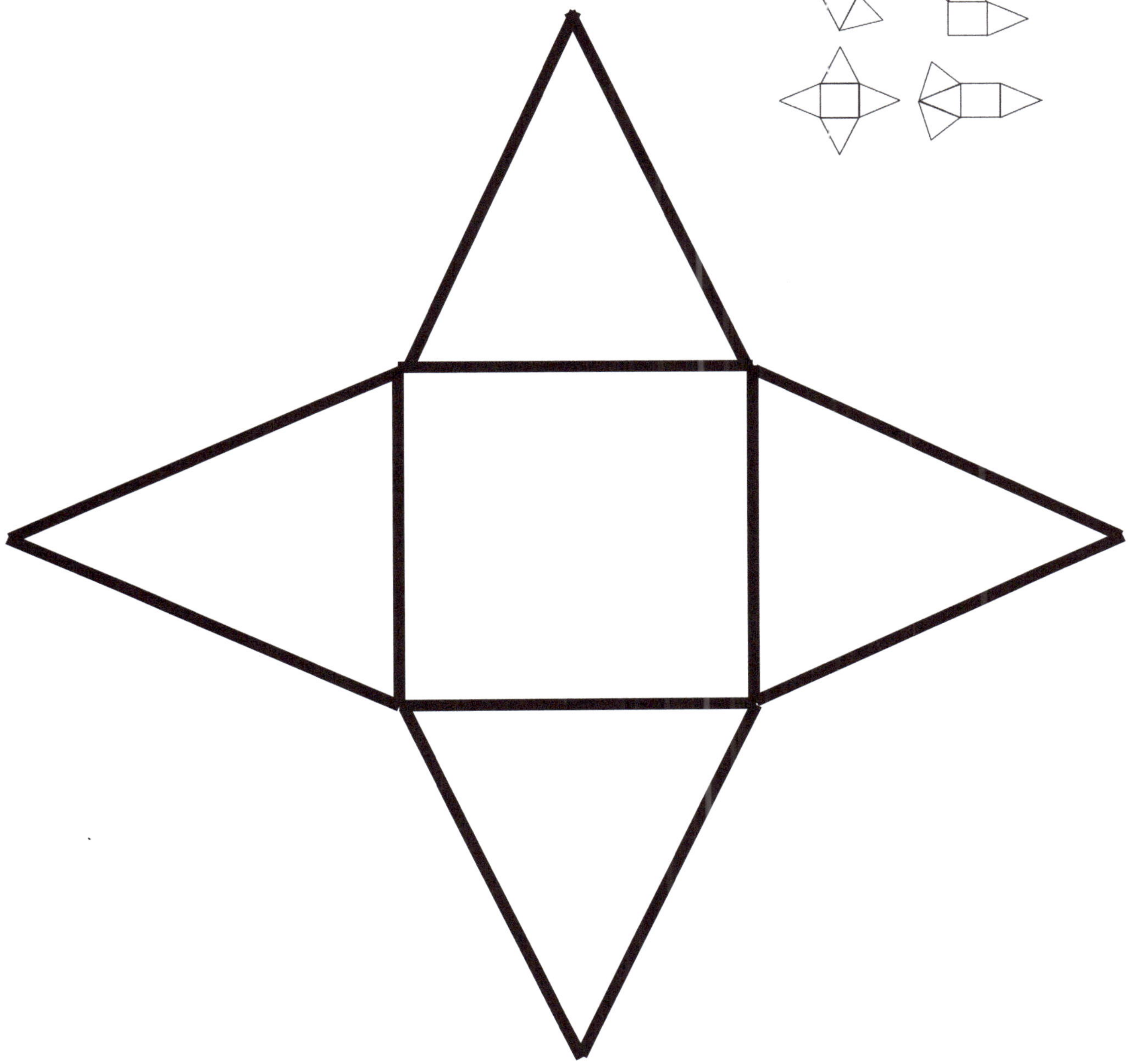

CUBE NET

INSTRUCTIONS: Print this net for use in a center or during STEMVESTIGATIONS.

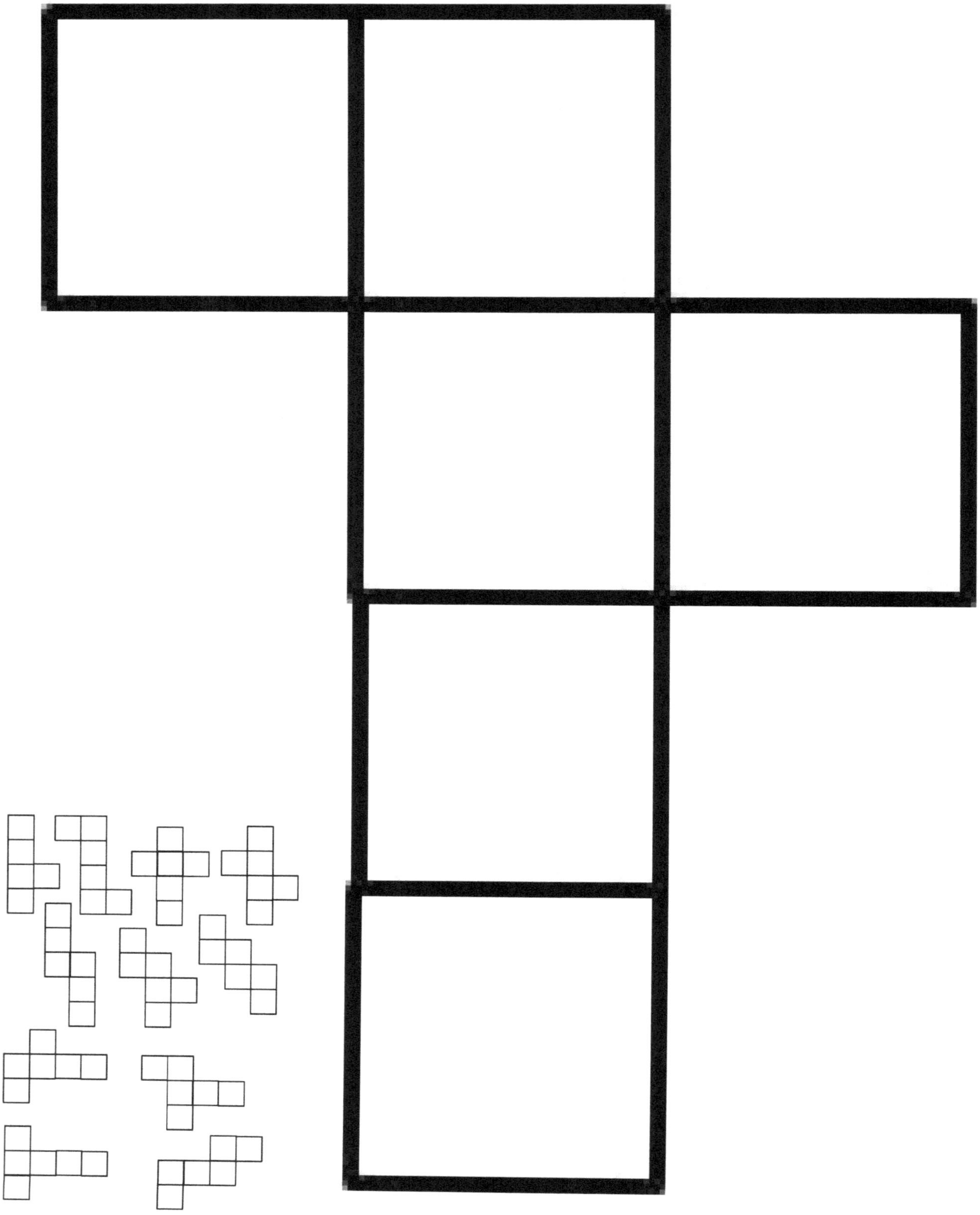

RHOMBI'S ADVENTURES IN 2D
RUBRIC: GENERAL

TASK **Students will use their knowledge of polygons to research animals and create unique habitats for shelter pets.**

	Content	Organization	Presentation
1	• Is well thought out and supports the solution to the challenge or question • Reflects application of critical thinking • Has clear goal that is related to the topic • Is accurate	• Information is clearly focused in an organized and thoughtful manner • Information is constructed in a logical pattern to support the solution	• Format enhances the content • habitat is appropriate for the animal • habitat utilizes polygons • habitat is solidly built
2	• Supports the solution • Has application of critical thinking that is apparent • Has no clear goal • Has some factual errors or inconsistencies (i.e. wheels do not rotate around axle)	• Project has a focus but might stray from it at times (more concerned with form than function) • Information appears to have a pattern, but the pattern is not consistently carried out in the project	• Format does not suit the content • habitat is vaguely appropriate for the animal • habitat uses polygons • habitat is wobbly
3	• Provides inconsistent information for solution • Has no apparent application of critical thinking • Has no clear goal • Has significant factual errors, misconceptions, or misinterpretations	• Content is unfocused and haphazard • Information does not support the solution to the challenge or question • Information has no apparent pattern	• Presentation appears sloppy and/or unfinished

Rhombi's ADVENTURES IN SPACE AND SHAPE

CONTENTS

RHOMBI'S ADVENTURES IN SPACE AND SHAPE
MATH MODELING IN K-2 SETTINGS

DEFINING THE MATH MODEL

Mathematics can be used to "model", or represent, how the real world works. A model is not the real thing, but it should be accurate enough to be useful. Math models connect abstract numbers to real experience. Math models save time, supplies, and aggravation. For example, Rhombi needs to build a full-scale cart. Since the class cannot build the full sized vehicle, it will develop a small scale version and extrapolate from there. They use their background knowledge to begin the math model.

MOVE FROM EXPERIENCE TO SYMBOLS

The study of science and math started in an effort to understand actions that could be observed directly. In order to make science and math a useful performing art, instruction must start with real experience and gradually build toward an abstract representation. Doing things in the reverse chronological order produces nearly useless results.

KNOW WHERE THE LESSON IS HEADED

Education separates experience from the learning process. We could learn a few things from our own history. Math development progressed along its slow, continuous track: caveman to calculus. Here, in the 21st century, we expect second graders to operate at abstract levels using a symbolic language that took mathematicians thousands of years to develop. Educators are asked to start with the algorithm -the abstraction- and work backwards to the concrete. We wonder why math is such a dreaded subject. Neither the teacher nor the student knows where the symbols are supposed to take them.

STEAM USES NATURAL EXTENSIONS

Abstract models of STEAM phenomena should grow as a natural extension of what is felt by direct experience. The abstraction must be an immediate and obvious representation of what students already know at every level. Drive a radio controlled car. Build the shapes as you memorize them. Investigate, collect data and analyze results. From reality to abstraction, STEAM learning needs more teachers who understand that doing more means going further, faster.

DEFINE YOUR MATHEMATICAL MODEL

Identify the problem, define the terms in your problem, & draw diagrams where appropriate. Begin with a simple model, stating assumptions that you make as you focus on particular aspects of the phenomenon. Identify important variables and constants. Determine how they relate to each other. Develop the equation(s) that express relationships between variables & constants.

ASSESS NUMERACY & CARDINALITY

Print rubric for your records. Perform assessment as early as possible.

STUDENT NAME	1	2	3	4	5	6	7	8	9

NOTE: All students will need to break amounts into smaller groups in order to recognize the cardi-nal numbers beyond "4." The brain naturally grasps groups of 1,2,3,4. It's why phone numbers are broken into smaller groups.

RHOMBI'S ADVENTURES IN SPACE AND SHAPE
SCOPE & SEQUENCE

Rhombi's Adventures in Geometry

TEACHER'S PRINTABLE UNIT PLANNING PAGE

2D SHAPES: UNIT ____________					
STEMVESTIGATION	DATE	TIME	WHAT I'VE GOT	WHAT I NEED	PLANNING NOTES
SCIENCE					
TECHNOLOGY					
ENGINEERING					
ARTS					
MATH					
CHALLENGE					

TEACHER'S PRINTABLE CHALLENGE PLANNING PAGE

2D SHAPES: UNIT ______					
STEMVESTIGATION	DATE	TIME	WHAT I'VE GOT	WHAT I NEED	PLANNING NOTES
CHALLENGE:					
BRAINSTORM & HYPOTHESIZE					
DESIGN					
BUILD					
TEST					
ANALYZE DATA					

RHOMBI'S ADVENTURES IN SPACE AND SHAPE
VOCABULARY

CIRCLE TIME	
anemometer	An instrument for measuring wind force and velocity.
ellipse	essentially, a circle or an oval
oval	egg-shaped
temperature	degree of hotness or coldness
weather	The state of the atmosphere at a given time and place, with respect to variables such as temperature, moisture, wind velocity
SQUARE ONE	
attribute	properties, qualities, features of a thing
magnifying glass	A lens or combination of lenses that enlarges the image of an object.
origami	The art or process, originating in Japan, of folding paper into shapes representing flowers and birds, for example.
tallies	marks that represent a given amount

VOCABULARY

WE LOVE TRIANGLES	
polygon	a plane figure bounded by three or more line segments
primary color	a color, as red, yellow, or blue, that in mixture yields other colors
scale model	an object, built to scale, that represents in detail another, often larger objects
secondary color	a color, as orange, green, or violet, produced by mixing two primary colors
TAKE FIVE	
innovate	to begin or create something new
optimize	to make as perfect or efficient as possible
pentagon	five sided plane figure / polygon with five edges and five corners (vertices)
RHOMBI'S BEST IN SHOW	
habitat	the environment in which an animal or plant normally lives or grows
animal shelter	facility that houses and disposes of homeless, lost, or abandoned animals

WORD WALL WORDS FOR MATH

The one who does all the work also does all the learning. Students should illustrate the words for a word wall. Just looking at the cute cards we all like to print and laminate is not particularly effective as a learning tool. While the student created cards take longer (and look significantly messier), the benefits are greater.

Count	Equal	Less
Zero	Addition	Taller
Greater than	Subtraction	Shorter
Less than	Ones	Longer
Higher	Tens	Larger
Lower	More	Smaller

Hour	Above	Circle
Minute	Below	Triangle
Seconds	Beside	Rectangle
Forward	In front of	Flat shape
Length	Behind	Solid shape
Weight	Next to	Corners
Height	Square	Sides

RHOMBI'S POLYGON ROUND-UP TEACHER'S BACKGROUND

GENERAL INFORMATION ON POLYGONS

A polygon is a geometric figure in two dimensions with three or more sides. The name comes from two Greek words, poly, meaning "many," and gon, meaning "angle." A polygon always has as many angles as it has sides. Polygons are named to indicate the number of sides or angles they contain. Thus, a hexagon has six (hexa- means "six") sides and six angles. The sum of a polygon's exterior angles is 360°...
To find the internal angles of a polygon use this formula:
Multiply 180 by (the number of sides minus 2). A hexagon has 6 sides. 6 - 2 = 4
Hexagon is 180° x 4 = 720°

POLYGON TERMINOLOGY

Parts and properties of polygons.

Side: Any one of the straight lines that make up the polygon.

Vertex: A point where any two of the sides of a polygon meet to form an angle.

Angle: A figure formed by the intersection of two sides.

Diagonal: A line that joins any two nonadjacent (not next to each other) vertices.

Perimeter: The sum of the length of all sides.

Area: The space enclosed within the polygon.

TYPES OF POLYGONS

Equilateral: A polygon in which all sides are equal in length.
Equiangular: A polygon in which all angles are the same size.

Regular: A polygon that is both equilateral and equiangular. This is the one most often used in schools.

EXAMPLES OF POLYGONS

The most common kinds of polygons include:

Parallelogram: A quadrilateral (four-sided figure) in which both pairs of sides are parallel and equal.

Rhombus: A parallelogram in which all four sides are equal.

Rectangle: A parallelogram in which all angles are right angles.

Square: A rectangle in which all four sides are equal.

SUM of INTERNAL ANGLES

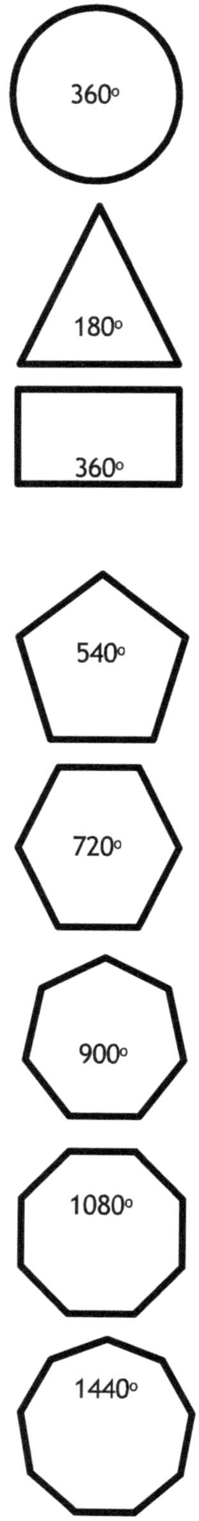

NGSS K 2

			ROUND	SQUARE	TRIANGLE	FIVES	MOVES	GIFTS
		Module	MODULE 1					
		STEMvestigations	ROUND	SQUARE	TRIANGLE	FIVES	MOVES	GIFTS
		Engineering Design Project	Cart	Tool	Bridge	Parachute	Move	
K-2	K-2.Engineering Design							
	K-2-ETS1-1	Ask questions, make observations, and gather information about a situation people want to change to define a simple problem that can be solved through the development of a new or improved object or tool.	x	x	x	x	x	x
	K-2-ETS1-2	Develop a simple sketch, drawing, or physical model to illustrate how the shape of an object helps it functionas needed to solve a given problem.	x	x	x	x	x	x
	K-2-ETS1-3	Analyze data from tests of two objects designed to solve the same problem to compare the strengths and	x	x	x	x	x	x
K	Forces and Interactions: Pushes and Pulls							
	K-PS2-1	Plan and conduct an investigation to compare the effects of different strengths or different directions of pushes and pulls on the motion of an object.	x		x	x	x	
	K-PS2-2	Analyze data to determine if a design solution works as intended to change the speed or direction of an object with a push or a pull.*	x			x	x	
K	Interdependent Relationships in Ecosystems: Animals, Plants, and Their Environment							
	K-LS1-1	Use observations to describe patterns of what plants and animals (including humans) need to survive					x	
	K-ESS2-2	Construct an argument supported by evidence for how plants and animals (including humans) can change the environment to meet their needs					x	
	K-ESS3-1	Use a model to represent the relationship between the needs of different plants or animals (including humans)		x	x		x	
	K-ESS3-3	Communicate solutions that will reduce the impact of humans on the land, water, air, and/or other living things in the local environment.		x	x		x	
K	Weather and Climate							
	K-PS3-1	Make observations to determine the effect of sunlight on Earth's surface.		x				
	K-PS3-2	Use tools and materials to design and build a structure that will reduce the warming effect of sunlight on an area.						
1	K-ESS2-1	Use and share observations of local weather conditions to describe patterns over time.	x	x		x	x	
	Waves: Light and Sound							
	1-PS4-1	Plan and conduct investigations to provide evidence that vibrating materials can make sound and that sound can make materials vibrate.						
	1-PS4-2	Make observations to construct an evidence-based account that objects can be seen only when illuminated.	x	x				
	1-PS4-3	Plan and conduct an investigation to determine the effect of placing objects made with different materials in the path of a beam of light.		x				
	1-PS4-4	Use tools and materials to design and build a device that uses light or sound to solve the problem of communicating over a distance.	x					

RHOMBI'S POLYGON ROUND-UP CIRCLE

ABOUT THE CIRCLE

- In the math world, circle refers to the boundary of the shape. "Disk" is used to refer to the whole shape, including the inside.
- A straight line from the center of a circle to the edge is called the radius.
- A straight line that passes from one side of a circle to the other through the center is called the diameter.
- The distance around the outside of a circle is called the circumference. All points on the edge of a circle are the same distance to the center.
- The value of Pi (π) to 2 decimal places is 3.14, it comes in handy when working out the circumference and area of a circle.
- Circles have a high level of symmetry.

RHOMBI'S POLYGON ROUND-UP
TRIANGLE

ABOUT THE TRIANGLE

- Triangles are polygons with the least possible number of sides (three).
- The three internal angles of a triangle always add to 180 degrees.
- An equilateral triangle has three sides of equal length and three equal angles.
- The longest side of a right angle triangle is called the hypotenuse, it is always found opposite the right angle.
- Trigonometry is the study of the relationship between the angles of triangles and their sides.
- Triangle shapes are often used in construction because of their great strength.

RHOMBI'S POLYGON ROUND-UP
RECTANGLE

ABOUT THE SQUARE

- A square is a polygon with 4 sides of equal length and 4 right angle corners (90 degree corners).
- Because it has 4 sides of equal length, a square is a regular quadrilateral.
- A square is also a rectangle with equal sides and a rhombus with right angles.
- The perimeter of a square is 4 times the length of one side.
- Opposite sides of a square are parallel.
- The internal angles of a square add to 360 degrees.
- A square has 4 lines of reflectional symmetry.

RHOMBI'S POLYGON ROUND-UP PENTAGON

ABOUT THE PENTAGON

- A pentagon is a 5 sided polygon with interior angles that add to 540 degrees.
- Regular pentagons have sides of equal length and interior angles of 108 degrees.
- The US Department of Defense headquarters is named 'the Pentagon'.
- The edible plant okra is shaped like a pentagon.

RHOMBI'S POLYGON ROUND-UP HEXAGON

ABOUT THE HEXAGON

- A hexagon is a 6 sided polygon with interior angles that add to 720 degrees.
- Regular hexagons have sides of equal length and interior angles of 120 degrees.
- Beehive cells are hexagonal.

RHOMBI'S POLYGON ROUND-UP HEPTAGON

ABOUT THE HEPTAGON

- A heptagon is a 7 sided polygon with interior angles that add to 900 degrees.
- Regular heptagons have sides of equal length and interior angles of 128.57 degrees.
- The British 50 and 20 pence coins are curved heptagons.

RHOMBI'S POLYGON ROUND-UP
OCTAGON

ABOUT THE OCTAGON

- An octagon is an 8 sided polygon with interior angles that add to 1080 degrees.
- Regular octagons have sides of equal length and interior angles of 135 degrees.

RHOMBI'S POLYGON ROUND-UP
NONAGON

ABOUT THE NONAGON

- A nonagon is a 9 sided polygon with interior angles that add to 1260 degrees.
- Regular nonagons have sides of equal length and interior angles of 140 degrees.

RHOMBI'S POLYGON ROUND-UP
DECAGON

ABOUT THE DECAGON

- A decagon is a 10 sided polygon with interior angles that add to 1440 degrees.
- Regular decagons have sides of equal length and interior angles of 144 degrees.

CUBE NET

INSTRUCTIONS: Print this net for use in a center or during STEMVESTIGATIONS.

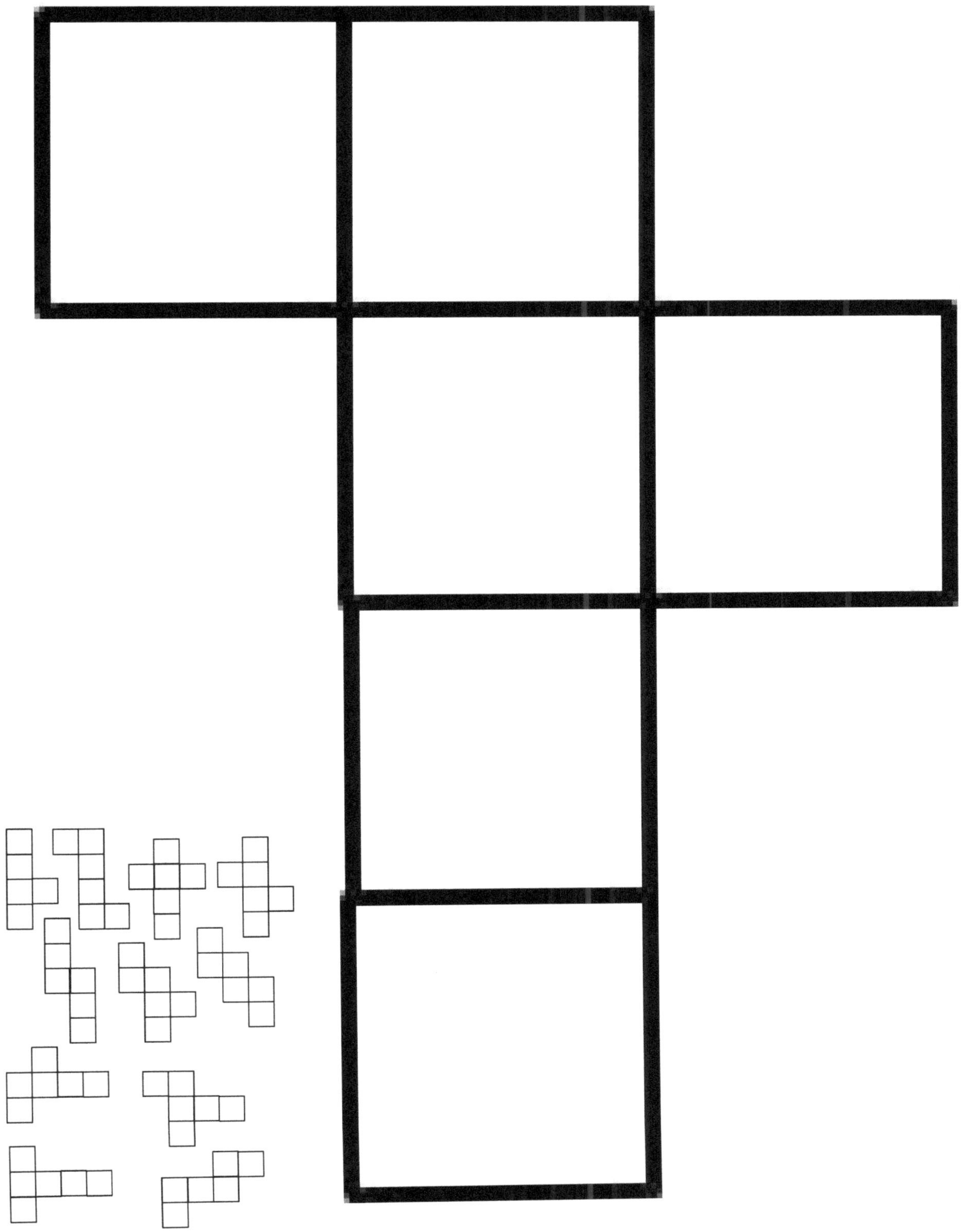

Rhombi's AADVENTURES IN SPACE AND SHAPE

PRE AND POST ASSESSMENT PAGES

RECOGNIZE SHAPES

Color the square.		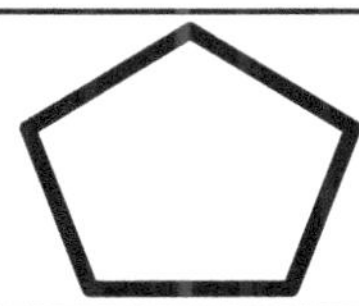		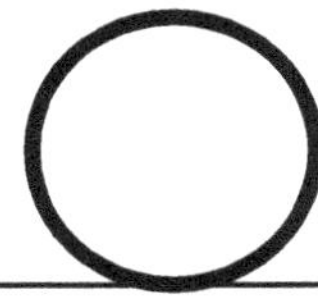
Color the triangle.		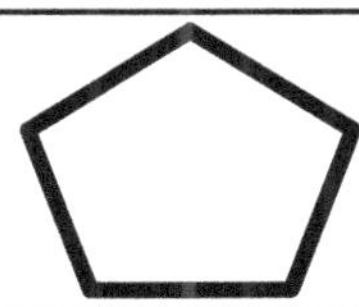		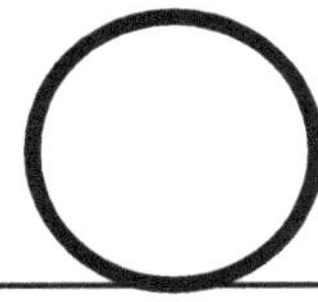
Color the rectangle.		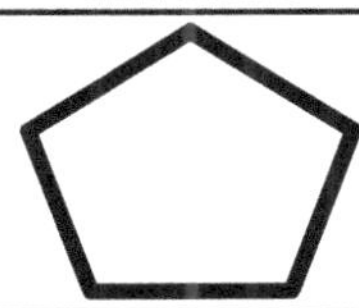		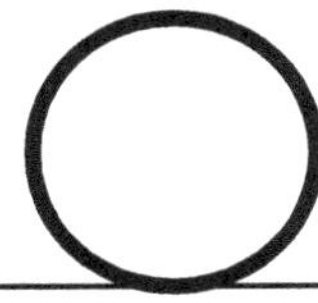
Color the circle.		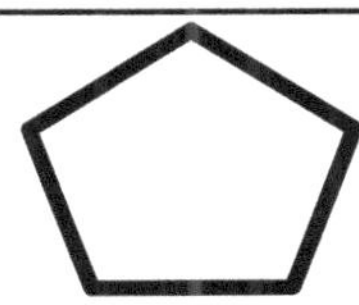		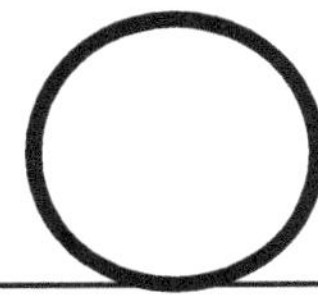
Color the pentagon.		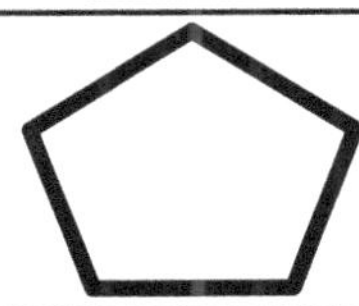		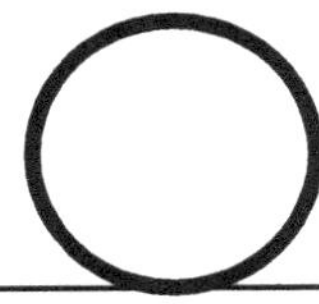

REPRODUCE SHAPES

Draw a square.	
Draw a triangle.	
Draw a rectangle.	
Draw a circle.	
Draw a pentagon.	

MOVE FROM 2D TO 3D

Mark the shape you see on a cube.

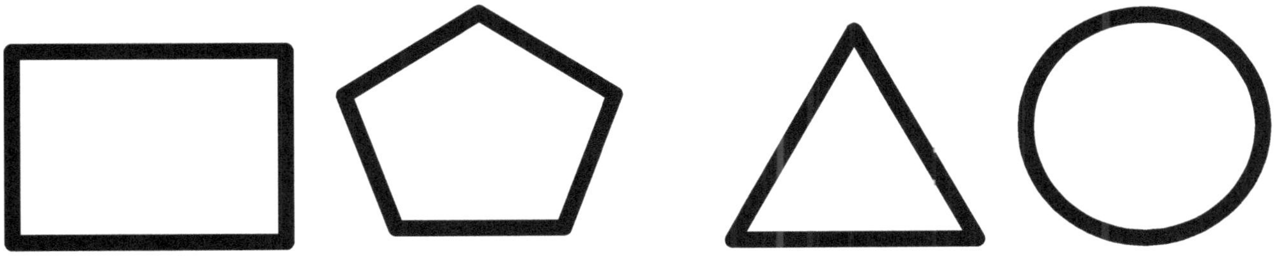

Mark the shape you see on a pyramid.

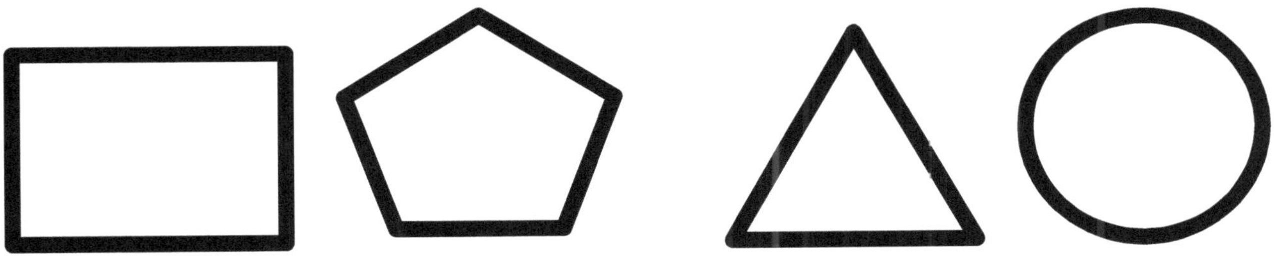

Mark the shape you see on a the end of a cylinder.

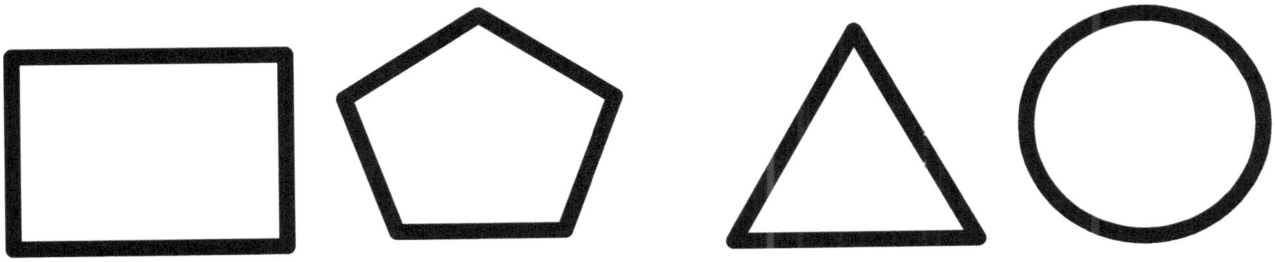

UNDERSTAND EQUALITY

Draw a line to break each shape into two equal parts.

Draw the other half of each shape.

RECOGNIZE SHAPES IN THE WORLD

Glue the cut shapes to this page to make a picture.

DO NOT COPY FOR STUDENT
one-on-one with teacher

Show the child 2 pennies. Ask, "How many objects are in my hand?"

Child immediately says, "2."Has solid sense of abstract number 2... Ready for discussions without visuals...

Child counts out loud and answers "2."Has basic understanding of abstract number 2... Ready for handouts...

Child points to each object, counting out loud and answers "2."
Needs concrete objects to count while transitioning to abstract number sense...

Repeat actions for numbers 3 and 4.

Recognizing the sum of a pile of objects beyond 4 is almost impossible. Try it yourself. You will reflexively break the larger number into two or more lesser groups before combining to say the greater number.

All students will break numbers beyond 4 into smaller "piles." Note which students count by ones beyond 4 and which students automatically do some mental addition to give you the whole number.

ASSESS OBJECT PERMANENCE

DO NOT COPY FOR STUDENT
one-on-one with teacher

Show the child your hand as you fold 2 pennies into your closed fist. Reopen the same hand. Ask. How many objects are in my hand?"

Child immediately says, "2."Has solid sense of object permanence for 2... Ready for discussions without visuals...

Child counts out loud and answers "2."Has basic understanding of object permanence for 2. Ready for handouts...

Child points to each object, counting out loud and answers "2."
Needs concrete objects to count while transitioning to object permanence...

Repeat actions for numbers 3 - 9.

Remember, in this assessment, your goal is to determine a student's grasp of object permanence
- NOT ability to count. The outcome is dependent on the student's ability to restate the cardinal number.

CCSS Mathematics Counting & Cardinality K

	Date Taught	Date Assessed
Know number names and the count sequence.		
I can count to 100 by ones and tens.		
I can count forward starting at any number I have learned.		
I can write numbers from 0 to 20.		
I can write a number to tell about a group of 0 to 20 things.		
Count to tell the number of objects.		
I can understand how number names go with counting things in the right order.		
I can name the number for each thing in a group as I count them.		
I can understand that the last thing I count tells the number of things in a group.		
I can understand that things in a group can be moved around and the total number will be the same.		
I can understand that the next number I say when I count means that there is one more.		
I can count up to 10 to tell how many things are in a group.		
I can count out a group of things when someone gives me any number from 1to 20.		
Compare numbers.		
I can use matching or counting to tell if a group of objects in one group is bigger, smaller, or the same as a group of objects in another group.		
I can compare two written numbers between 1 and 10.		

CCSS Mathematics Operations & Algebraic Thinking K			
Indicator		**Date Taught**	**Date Assessed**
Understand addition, and understand subtraction.			
CCSS.MATH.CONTENT.K.OA.A.1	I use what makes sense to me to show that I know how to add.		
CCSS.MATH.CONTENT.K.OA.A.1	I use what makes sense to me to show that I know how to subtract.		
CCSS.MATH.CONTENT.K.OA.A.2	I use objects or drawings to show that I can solve addition word problems up to 10.		
CCSS.MATH.CONTENT.K.OA.A.2	I use objects or drawings to show that I can solve subtraction word problems up to 10.		
CCSS.MATH.CONTENT.K.OA.A.3	I take apart any number from 1to 10 to show that I understand that number. (5 = 2 + 3)		
CCSS.MATH.CONTENT.K.OA.A.4	I take any number from 1to 9 and show what I need to add to it to make 10.		
CCSS.MATH.CONTENT.K.OA.A.5	I add numbers within 5.		
CCSS.MATH.CONTENT.K.OA.A.5	I subtract numbers within 5.		

CCSS Mathematics Base Ten K			
Indicator		**Date Taught**	**Date Assessed**
Work with numbers 11-19 to gain foundations for place value.			
CCSS.MATH.CONTENT.K.NBT.A.1	1 make and take apart numbers from 11to 19 by telling how many tens and ones are in the number.		
CCSS.MATH.CONTENT.K.NBT.A.11 can show how many tens and ones in numbers from 11to 19 by drawing a picture or writing a number sentence.	1 show how many tens and ones in numbers from 11to 19 by drawing a picture or writing a number sentence.		

CCSS Mathematics Measurement & Data K			
Indicator		Date Taught	Date Assessed
Describe and compare measurable attributes.			
CCSS.MATH.CONTENT.K.MD.A.1I			
CCSS.MATH.CONTENT.K.MD.A.2			
Classify objects and count the number of objects in each category.			
CCSS.MATH.CONTENT.K.MD.A.3			
CCSS.MATH.CONTENT.K.MD.A.3			

Indicator		Date Taught	Date Assessed
Identify and describe shapes.			
CCSS.MATH .CONTENT.K.G .A.1 I can name and tell about shapes I see around me.	I name and tell about shapes I see around me.		
CCSS.MATH .CONTENT.K.G .A.1 I can tell where I see shapes by using words like: above, below, beside, in front of, behind and next to.	I tell where I see shapes by using words like: above, below, beside, in front of, behind and next to...		
CCSS.MATH .CONTENT.K.G .A.2	I name shapes no matter how big they are or which way they are turned.		
CCSS.MATH .CONTENT.K.G .A.3	I tell if a shape is two-dimensional (flat) or three-dimensional (solid).		
Analyze, compare, create and compose shapes.			
CCSS.MATH .CONTENT.K.G .A.4	I think about and compare two-dimensional and three-dimensional shapes.		
CCSS.MATH .CONTENT.K.G .A.5	I make shapes by drawing them or by using things like sticks and clay.		
CCSS.MATH .CONTENT.K.G .A.6	I use simple shapes to make larger shapes.		

CCSS Reading Foundations K			
Indicator		Date Taught	Date Assessed
Print Concepts			
CCSS.ELA-Literacy.RF.K.l	I show that I know how books should be read.		
CCSS.ELA-Literacy.RF.K.l.A	I read the words in a book in the right order.		
CCSS.ELA-Literacy.RF.K.l.B	I understand that words I say can be written using letters in a certain order.		
CCSS.ELA-Literacy.RF.K.l.C	I understand that words have spaces between them.		
CCSS.ELA-Literacy.RF.K.l.D	I name all of my upper and lower case letters in the alphabet.		
CCSS.ELA-Literacy.RF.K.2	I show that I know how words and their parts go together.		
CCSS.ELA-Literacy.RF.K.2.A	I recognize and make rhyming words.		
CCSS.ELA-Literacy.RF.K.2.B	I count and divide words into syllables.		
CCSS.ELA-Literacy.RF.K.2.C	I blend and take apart the beginning sounds and ending parts of one-syllable words.		
CCSS.ELA-Literacy.RF.K.2.0	I find and say the beginning, middle and last sound in simple words.		
CCSS.ELA-Literacy.RF.K.2.E	I make new words by changing a consonant or a vowel sound in a word I already know.		
Phonics and Word Recognition			
CCSS.ELA-Literacy.RF.K.3	I show what I have learned about letters and sounds by reading words.		
CCSS.ELA-Literacy.RF.K.3.A	I say the most common sound for each consonant in the alphabet.		
CCSS.ELA-Literacy.RF.K.3.B	I match long and short vowel sounds with the letters that go with them.		
CCSS.ELA-Literacy.RF.K.3.C	I read common sight words.		
CCSS.ELA-Literacy.RF.K.3.0	I tell the difference between and read similar words by looking at the letters that are different.		
Fluency			
CCSS.ELA-Literacy.RF.K.4	I read and understand books at my level well.		

CCSS Reading	Literature K		
Indicator		**Date Taught**	**Date Assessed**
Key Ideas and Details			
CCSS.ELA-LITERACY.RL.K.1	I ask and answer questions about important details in stories.		
CCSS.ELA-LITERACY.RL.K.2	I retell a story I know using important details from the story.		
CCSS.ELA-LITERACY.RL.K.3	I know the characters, setting and what happens in stories.		
Character and Structure			
CCSS.ELA-LITERACY.RL.K.4	I ask and answer questions about new words in stories.		
CCSS.ELA-LITERACY.RL.K.5	I know difference between the different kinds of fiction I read.		
CCSS.ELA-LITERACY.RL.K.6	I name a book's author and illustrator and know what their jobs are.		
Integration of Knowledge and Ideas			
CCSS.ELA-LITERACY.RL.K.7	I know how the words and pictures go together in stories.		
CCSS.ELA-LITERACY.RL.K.9	I know what is the same and different about the actions of characters in stories I know.		
Range of Reading and Level of Text Complexity			
CCSS.ELA-LITERACY.RL.K.10	I am an important part of fiction reading activities in my classroom.		

CCSS Reading	Writing K		
Indicator		**Date Taught**	**Date Assessed**
Text Types and Purposes			
CCSS.ELA-LITERACY.W.K.1	I draw or write to help share what I think.		
CCSS.ELA-LITERACY.W.K.2	I draw or write to explain a topic.		
CCSS.ELA-LITERACY.W.K.3	I draw or write to tell an organized story about something that has happened.		
Production and Distribution of Writing			
CCSS.ELA-LITERACY.W.K.5	I listen to my friends' ideas to help add details to my stories.		
CCSS.ELA-LITERACY.W.K.6	I use technology to share my ideas.		
Research to Build and Present Knowledge			
CCSS.ELA-LITERACY.W.K.7	I help my class learn about a subject, and then I write about it.		

Indicator		Date Taugh t	Date Assessed
Key Ideas and Details			
CCSS.ELA-LITERACY.RI.K.l	I ask and answer questions about important details in nonfic-tion books.		
CCSS.ELA-LITERACY.RI.K.2	I name the main topic and important details i n nonfiction books.		
CCSS.ELA-LITERACY.RI.K.3	I tell how people, events or ideas are connected.		
Craft and Structure			
CCSS.ELA-LITERACY.RI.K.4	I ask and answer questions about new words in nonfiction books.		
CCSS.ELA-LITERACY.RI.K.S	I find the front cover, back cover and title page in nonfiction books.		
CCSS.ELA-LITERACY.RI.K.6 I	I name the main topic and important details i n nonfiction books.		
Integration of Knowledge and Ideas			
CCSS.ELA-LITERACY.RI.K.7	I tell how the words and pictures go together in nonfiction books.		
CCSS.ELA-LITERACY.RI.K.8	I find the reasons an author gives to make the information more clear.		
CCSS.ELA-LITERACY.RI.K.9	I can tell how two nonfiction books about the same thing a re alike and different.		
Range of Reading and Level of Text Complexity			
CCSS.ELA-LITERACY.RI.K.lO	I am an important part of nonfiction reading activities in my classroom.		

Indicator		Date Taught	Date Assessed
Comprehension and Collaboration			
CCSS.ELA-LITERACY.SL.K.1	I show that I know how to have good conversations with my friends and teachers.		
CCSS.ELA-LITERACY.SL.K.1.A	I listen and take turns in conversations.		
CCSS.ELA-LITERACY.SL.K.1.B	I have long conversations with another person.		
CCSS.ELA-LITERACY.SL.K.2	I can talk about the important ideas in a story.		
CCSS.ELA-LITERACY.SL.K.3	I ask and answer questions about what I have heard.		
Presentation of Knowledge and Ideas			
CCSS.ELA-LITERACY.SL.K.4	I use details when I tell about people, pl aces and things.		
CCSS.ELA-LITERACY.SL.K.5	I draw to add details to what I share.		
CCSS.ELA-LITERACY.SL.K.6	I speak and share my ideas clearly.		

CCSS Reading	Language K		
Indicator		Date Taught	Date Assessed
Conventions of Standard English			
CCSS.ELA-LITERACY.L.K.l	I show that I know how to use words correctly when I write and speak.		
CCSS.ELA-LITERACY.L.K..Al	I print lots of upper and lowercase letters.		
CCSS.ELA-LITERACY.L.K..Bl	I use nouns (words that name) and verbs (action words).		
CCSS.ELA-LITERACY.L.K..Cl	I make nouns plural (more than one) by adding "s" or "es" to the end.		
CCSS.ELA-LITERACY.L.K..Ol	I understand and use question words. (who, what, where, when, why, how)		
CCSS.ELA-LITERACY.L.K..El	I use common prepositions (to, from, in, out, on, off, for, of, by, with).		
CCSS.ELA-LITERACY.L.K..Fl	I create longer complete sentences with my class.		
CCSS.ELA-LITERACY.L.K.2	I show that I know how to write sentences correctly.		
CCSS.ELA-LITERACY.L.K.2.A	I capitalize the first word in a sentence.		
CCSS.ELA-LITERACY.L.K.2.A	I capitalize the word "1".		
CCSS.ELA-LITERACY.L.K.2.B	I find and name punctuation at the end of a sentence.		
CCSS.ELA-UTERACY.L.K.2.C \	I write a letter or letters for most consonant sounds.		
CCSS.ELA-LITERACY.L.K.2.C	I write a letter or letters for most short vowel sounds.		
CCSS.ELA-LITERACY.L.K.2.0	I use what I know about letters and sounds to spell easy words.		
Vocabulary Acquisition and Use			
CCSS.ELA-LITERACY.L.K.4	I figure out what words mean by thinking about what I have read.		
CCSS.ELA-LITERACY.L.K.4.A	I find new meanings for words I already know and use them correctly.		
CCSS.ELA-LITERACY.L.K.4.B	I use beginnings and endings of words to help me figure out what they mean (-ed, -s, re-, un-, pre-, -ful, -less).		
CCSS.ELA-LITERACY.L.K.S	I figure out how words are related. I figure out how meanings might be alike.		
CCSS.ELA-LITERACY.L.K.S.A	I sort things into groups. I use groups to understand things better.		
CCSS.ELA-LITERACY.L.K.S.B	I match some action verbs and adjectives with their opposites.		
CCSS.ELA-LITERACY.L.K.S.C	I tell how words are used in real-life.		
CCSS.ELA-LITERACY.L.K.S.O	I tell the difference between verbs (action words) that are almost alike.		
CCSS.ELA-LITERACY.L.K.6	I use new words in different ways to show that I know what they mean.		

RHOMBI'S ADVENTURES IN 2D
S.T.E.A.M. REFERENCE SHELF

"Circle."Wikipedia, the Free Encyclopedia. Wikipedia Foundation, Inc, n.d. Web. 7 July 2014.

Dehaene, Stanislas. The Number Sense: How the Mind Creates Mathematics. New York: Oxford UP, 1997. Print.

Devlin, Keith J. The Math Gene: How Mathematical Thinking Evolved and Why Numbers Are Like Gossip. New York: Basic Books, 2000. Print.

Ferris Bueller's Day Off. Dir. John Hughes. Perf. Matthew Broderick, Mia Sara, Alan Ruck, Jennifer Grey, and Jeffrey Jones. Paramount, 1986. Film.

Feynman, Richard P, and Jeffrey Robbins. The Pleasure of Finding Things Out: The Best Short Works of Richard P. Feynman. Cambridge: Perseus Books, 1999. Print.

Furoy, Michael, Rosemary T. Wong, and Harry K. Wong. The Effective Teacher. Mountain View: Harry K. Wong Publications, Inc., 2009. Print.

Ifrah, Georges. From One to Zero: A Universal History of Numbers. New York: Viking, 1985. Print.

Lee, Joon S., and Herbert P. Ginsburg. "Preschool Teachers' Beliefs About Appropriate Early Literacy and Mathematics Education for Low and Middle-Socioeconomic Status Children." Early Education and Development 18.1 (2007): 111-143. Print.

Ma, Liping. Knowing and Teaching Elementary Mathematics: Teachers' Understanding of Fundamental Mathematics in China and the United States. Hillsdale, NJ: Lawrence Erlbaum Associates, 1999. Print.

Mathemagic. Chicago: World Book—Childcraft International, Inc., 1980. Print.

McLeish, John. The Story of Numbers. New York: Fawcett Columbine, 1994. Print,

Next Generation Science Standards. N.p., n.d. Web. 7 July 2014.

Wassily Kandinsky - biography, paintings, books. N.p., n.d. Web. 7 July 2014.

"Inquiry Based Science: What Does It Look Like??" Connect Magazine. 1995: 13. Print.

Visit the International STEM League to learn more.
www.insl.org

www.ingramcontent.com/pod-product-compliance
Lightning Source LLC
LaVergne TN
LVHW071346210926
848486LV00003B/236
* 9 7 8 1 9 4 2 3 5 7 2 3 0 *